W. D. Ambrosch A. Maher B. Sasscer (Eds.)

The Intelligent Network

A Joint Study by Bell Atlantic,
IBM and Siemens

With 48 Figures

Springer-Verlag
Berlin Heidelberg New York
London Paris Tokyo

Wolf-Dietrich Ambrosch
IBM Deutschland GmbH
Pascalstrasse 100
D-7000 Stuttgart 80

Anthony Maher
Siemens AG
Wittelsbacher Platz
D-8000 München 1

Barry Sasscer
Bell Atlantic International
1300 N., 17th Street, Suite 1510
Arlington, VA 22209, USA

ISBN-13: 978-3-540-50897-7   e-ISBN-13: 978-3-642-46663-2
DOI: 10.1007/978-3-642-46663-2

2145/3140-54321 – Printed on acid-free paper

Preface

This report examines the findings of a study by Bell Atlantic, IBM, and Siemens which investigated the role of the Intelligent Network in telecommunications. It considers current trends and future developments, on a national and international level.

This report defines and describes the concept of the Intelligent Network and considers the operating requirements (including the hardware and software) and the types of service a network user can expect.

Concepts, definitions, and terminology reflect the status of the Intelligent Network in 1988. Actual implementation may differ due to the constantly changing environment, new requirements, and experiences with existing solutions.

The report is divided into six parts:

Part 1 introduces the concept of the Intelligent Network, and describes elements common to all IN services. It considers the application program and network management requirements, and provides examples of the hardware and software proposed for implementing the network.

Parts 2 through 6 include detailed descriptions of specific IN services. In each part there is a functional service description and an application description.

The reader is expected to have a general knowledge and understanding of existing telecommunication networks.

A bibliography, glossary, and all appendices referred to in this report are contained in the back of the document.

Contents

Introduction 1

Part 1. **Overview of the Intelligent Network** 3

Chapter 1. **The Intelligent Network** 5
1.1   IN Architecture and Capabilities 5
1.1.1   IN Goals 6
1.1.2   IN Technical Overview 8
1.2   IN Definition 9
1.2.1   IN Elements 9
1.2.2   An IN Service Example 12
1.3   IN-User Programmability 14
1.4   IN Introduction Scenarios 15
1.4.1   The Model 15
1.4.2   Introduction Scenarios 17
1.5   IN Architecture Validation 19

Chapter 2. **Functional Characteristics Common to Selected IN Services** 22
2.1   Overview 22
2.2   Methodology 22
2.3   Summary of Findings 23
2.4   Standards 24
2.5   Service Interaction 24
2.5.1   Service User 24
2.5.2   Service Subscriber 25
2.5.3   SMS Access Instrument 25
2.5.4   Network Operator 26
2.6   Billing 27
2.6.1   SSP 27
2.6.2   SCP 27
2.6.3   SMS 27
2.7   Service Logic 28
2.8   Databases 31
2.8.1   Size 31

2.8.2   Database Integrity Requirements   32
2.8.3   Location of Databases   32
2.8.4   Database Administration   33
2.9   Transaction Rate Estimates   34
2.9.1   Methodology   34
2.9.2   Transactions Between SMS and its Users   35
2.9.3   SMS-SCP Transaction Rates   35
2.9.4   SCP- SSP Transaction Rates   35
2.10   Measurement Requirements   36
2.10.1   SSP   36
2.10.2   SCP   38
2.10.3   SMS   39
2.11   Performance and Dynamic Requirements   39
2.11.1   Response Time   39
2.11.2   Response Times for IN Elements   40
2.11.3   Overload Handling   43
2.12   Availability Requirements and Objectives   44
2.13   National Considerations   45
2.14   Future Considerations   45

Chapter 3. **Network and Service Management**   47
3.1   Overview   47
3.2   Methodology   48
3.2.1   Summary of Findings   48
3.3   NSMS Requirements   49
3.3.1   NSMS Global Qualities   49
3.3.2   Basic Operations, Administration and Management Functions . .   49
3.3.3   Network Management Users   52
3.4   Switching Layer and the OSS—an Example   53
3.5   Signaling Network and the SEAS   54
3.5.1   Management Functions Related to the Signaling Network   54
3.6   IN-Services and the SMS   57
3.6.1   Service Management System Overview   58
3.6.2   SMS Function Description   59
3.7   Administrative System Functions and IN Requirements   62
3.7.1   IN-Wide Service Management Examples   62
3.8   Trends   63
3.8.1   Evolving Operations Architecture   63

Chapter 4. **Network Components**   66
4.1   Main Objectives of the IN Architecture   66
4.2   Network Topology   66
4.2.1   Performance and Throughput Requirements   67
4.2.2   Availability Requirements   68
4.2.3   IN Architecture   69
4.3   Service Control Point (SCP)   73

4.3.1	Requirements	. . . .	73
4.3.2	SCP Architecture	. . . .	76
4.4	Signaling Transfer Point (STP)	. . . .	87
4.4.1	Functional STP Requirements	. . . .	88
4.4.2	CCS7 Network Architecture and IN	. . . . . .	88
4.5	Service Switching Point (SSP)	. . . .	94
4.6	Future Considerations	. . . .	100

Part 2.   Green Number Service 103

Chapter 5.	**GNS Service Description**	. . . .	104
5.1	Overview	. . . .	104
5.2	Functional Description	. . . .	105
5.3	Standards	. . . .	107
5.4	Service Interaction	. . . .	107
5.5	Billing	. . . .	108
5.6	Service Logic	. . . .	109
5.6.1	Distribution	. . . .	109
5.6.2	Functional Flow	. . . .	109
5.7	Traffic Measurement Requirements	. . . .	110
5.8	Dynamic Requirements and Performance	. . . . .	112
5.9	National Dependencies	. . . .	112
5.10	Future Considerations	. . . .	113

Chapter 6.	**GNS Application Description**	. . . .	114
6.1	Functional Requirements and Allocation	. . . . .	114
6.2	Functional Units in the EWSD SSP	. . . .	117
6.3	Administrative Units in SSP	. . . .	120
6.4	Functional Units in SCP	. . . .	121
6.5	Administrative Units in SCP	. . . .	129

Part 3.   Alternate Billing Service 133

Chapter 7.	**ABS Service Description**	. . . .	134
7.1	Overview	. . . .	134
7.2	Functional Description	. . . .	135
7.3	Standards	. . . .	137
7.4	Service Interaction	. . . .	137
7.5	Billing	. . . .	138
7.6	Service Logic	. . . .	139
7.6.1	Distribution	. . . .	139
7.6.2	Functional Flow	. . . .	140
7.7	Traffic Measurements Requirements	. . . .	143
7.8	Dynamic Requirements and Performance	. . . . .	144

7.9	National Dependencies	144
7.10	Future Considerations	144

Chapter 8.   ABS Application Description 146

8.1	Functional Requirements and Allocation	146
8.2	Functional Units in SSP	149
8.3	Administrative Units in SSP	152
8.4	Functional Units in SCP	153
8.5	Administrative Units in SCP	157

Part 4.   Emergency Response Service 161

Chapter 9.   ERS Service Description 162

9.1	Overview	162
9.2	Functional Description	164
9.3	Standards	166
9.4	Service Interaction	167
9.5	Billing	169
9.6	Service Logic	169
9.6.1	Distribution	169
9.6.2	Functional Flow	170
9.7	Traffic Measurement Requirements	172
9.8	Dynamic Requirements and Performance	172
9.9	National Dependencies	172
9.10	Future Considerations	176

Chapter 10.   ERS Application Description 178

10.1	Functional Requirements and Allocation	178
10.2	Functional Units in SSP	181
10.3	Administrative Units in SSP	184
10.4	Functional Units in SCP	184
10.5	Administrative Units in SCP	194

Part 5.   Private Virtual Network 199

Chapter 11.   PVN Service Description 200

11.1	Overview	200
11.2	Functional Description	202
11.3	Standards	215
11.4	Service Interaction	216
11.5	Billing	217
11.6	Service Logic	218
11.6.1	Distribution	218
11.6.2	Functional Flow	221

11.7   Traffic Measurement Requirements 226
11.8   Dynamic Requirements and Performance   228

Chapter 12.   PVN Application Description   230
12.1   Functional Requirements and Allocation   230
12.2   Functional Units in SSP   241
12.3   Administrative Units in SSP   248
12.4   Functional Units in SCP   250
12.5   Administrative Units in SCP   255

Part 6.   Area Wide Centrex   257

Chapter 13.   AWC Service Description   258
13.1   Overview 258
13.2   Functional Description   259
13.3   Standards   263
13.4   Service Interaction   263
13.5   Billing   265
13.6   Service Logic   266
13.6.1   Distribution   266
13.6.2   Functional Flow   267
13.7   Traffic Measurement Requirements 278
13.8   Dynamic Requirements and Performance   280
13.9   National Dependencies   280
13.10   Future Considerations   280

Appendix A. Supplementary Services   281

**Appendix B. Bellcore Preliminary-Defined Functional
                       Components and Requests**   283

Glossary   285

Bibliography   293

List of Figures

1. Access Lines and Corporate Revenue, 1985[2] 7
2. IN Components 8
3. IN Elements 10
4. Non-IN GNS Service 12
5. IN-Based Green Number Service 13
6. IN-Pay-Per-View 20
7. Database Distribution 33
8. STP Message Delay[13] 40
9. SCP Response Time of SMS Generic Messages 41
10. SCP Response Time for SSP Messages (Service Specific) 42
11. Operation Systems within the United States' Intelligent Network
    Operations 50
12. Network User Categories 51
13. Operations System Users 52
14. SEAS Control of CCS7 Network 55
15. SMS Overview 58
16. Protocol Architecture in OA&M Networks 65
17. IN User Interfaces 69
18. IN Components and their Related Interfaces 71
19. Basic Data Flow Between an SSP and SCP (for GNS Without Prompting
    Option) 72
20. SCP Major Components 77
21. SCP Software Components 80
22. General STP Architecture for the EWSD 90
23. Possible IN CCS7 Architecture for the EWSD 91
24. Architecture for the EWSD 92
25. Common Channel Network Control for the EWSD 93
26. CP113 Coordination Processor for the EWSD 94
27. The Operating System 95
28. SSP Components 96
29. EWSD System Structure and Information and Message Flow for the GNS 99
30. GNS Functional Flow—GNS without Prompting Option 111
31. Functional Flow of ABS with Collect Call, Third Party Billing, and
    Manual Calling Card Billing 141
32. Functional Flow of ABS with Automatic Calling Card Billing 142

33. ERS Version 1, with Access to the SCP Database for ALI 171
34. ERS Version 2, with Access to the DAS Database for ALI 174
35. ERS Version 3, ALI Retrieved from SCP at Call Set-UP 177
36. SCP Functional Units and Application Platform Interfaces 185
37. SCP Administrative Units and Application Platform Interfaces 195
38. Standard PVN Transaction, Type 1 223
39. Standard PVN with Resource Counters, Type 2 224
40. PVN with Remote Access Using PIN, Type 3 225
41. PVN Application Functional and Administrative Units 251
42. AWC Intercom Dialing Functional Flow 269
43. AWC Call Forwarding Functional Flow 271
44. Speed Calling Functional Flow 272
45. MLHG Distribution in the SSP Functional Flow 274
46. MLHG Distribution in the SCP Functional Flow 275
47. Call Extension Functional Flow 277
48. Night Service Functional Flow 279

List of Tables

1. IN Introduction Evolution Scenario 18
2. TCAP Messages Between SSP and SCP 29
3. Generic Messages Between SCP and SMS[10] 30
4. Trigger Table Example[17] 46
5. Centrex and Possible Area Wide Centrex Features 260

Introduction

In the short time since its creation, the Intelligent Network (IN) has captured the imagination of the telecommunications industry. This fascination has been inspired by the promises inherent in this network service control architecture. The current excitement is based on the IN's functional architecture, its benefits, and its evolution.

There is already considerable knowledge of these attributes, however, this knowledge is located in many evolving documents and is still incomplete. The body of knowledge needs to be, and is being enlarged. Therefore, the telecommunications industry is steadily and probingly ascending the IN learning curve.

The cause and effect relationship between curiosity (business and technical) and information acquisition in an atmosphere of worldwide technological advances is propelling the IN onward. The goal of the Bell Atlantic, IBM, and Siemens joint study is to contribute to the understanding and realization of the IN architecture and its benefits. In the process, all study participants have enriched their own understanding of the IN through the exchange of ideas that is intrinsic in study interaction. The collaborative effort involved three corporations with expertise in networking, information processing, and telecommunications and produced a unique synthesis of analysis, experience and perspective.

The results of the detailed studies have been integrated and streamlined into this report which focuses on the important aspects of the IN: the architecture, services, and solutions.

In both the study methodology and the consolidation of the results into this document, emphasis has been placed on the current IN factors, however, IN's evolutionary phases are also considered.

Based on their available resources, the study teams set out to establish:

- A consolidated common understanding (among the study participants) of the IN concept, and whether its architecture is applicable worldwide.
- A common technical ground among the different corporations, with respect to elements of the IN architecture.
- Implementation scenarios for the IN elements, to determine the types of solutions available with the IN, and the limitations on these solutions.
- A general functional understanding of the important IN service characteristics such as: a description of what each service provides, the functional distribution of a service among the IN elements, network and service user interaction, and

functional aspects common to the services selected for detailed investigation. These services are the Green Number Service (GNS), Emergency Response Service (ERS), Alternate Billing Service (ABS), Private Virtual Network (PVN), and Area Wide Centrex (AWC).

Bell Atlantic, IBM, and Siemens developed a methodology to determine the transaction rates and storage capacity requirements for IN elements. This is based on demographic data, interviews and knowledge of telecommunications markets worldwide.

An appreciation of the future of the Intelligent Network is extremely important to a complete understanding of the current IN. The IN is becoming increasingly sophisticated with respect to defining the software building blocks (functional components) from which the services of the future will be constructed. These functional components (FCs) will not only be standardized products among telecommunications manufacturers, but will also be combinable to produce new services flexibly and efficiently. The collection of functional components will be analogous to the standard instruction sets in microprocessors, in which individual instructions can be combined in unique ways to produce many new software programs.

Telecom service creation will then be possible by linking the required FCs. Thus, the opportunity to conceptualize and implement a service will be available to other parties in the telecom industry instead of (as is currently the case) only the equipment suppliers. One of these parties will be the PTTs themselves. Therefore, since the PTTs are customers from the suppliers' perspective, this new service creation capability has often been referred to as "customer programmability". The future will also bring new and universally accepted standards in the areas of the IN interface protocols and service definitions. The achievement of standards is essential to IN progress.

It is also important to realize that the IN service control architecture and the ISDN access architecture will synergistically bring added opportunities. New service attributes will be provided which benefit the user and enhance service value through the combination of IN and ISDN. It will be seen that these two network architectures are complementary and that the whole is much greater than the sum of the parts.

Part 1. Overview of the Intelligent Network

This part defines and describes the Intelligent Network.

Chapter 1, "The Intelligent Network" introduces the concept, explains the origins of the IN, and gives a technical overview of network and service implementation. It describes IN introduction and evolution scenarios, which include the following environments for IN introduction:

- Conventional (analog subscriber) networks
- High density ISDN networks
- IN growth coincident with ISDN
- Deployment and penetration of ISDN or CCS7[1].

This chapter also presents a validation of the IN architecture and its capabilities by applying it to a specific service.

Chapter 2, "Functional Characteristics Common to Selected IN Services" describes the functional characteristics common to selected IN services. The information in this chapter applies to all selected services, and service-specific details are included in the individual service descriptions (Parts 2 through 6).

This chapter examines the standards to be applied to each service and how the service will be used. The functional characteristics described include billing, service logic, databases, traffic measurement, and performance requirements.

Chapter 3, "Network and Service Management" considers aspects of network and service management. It considers the basic Operations, Administration, and Management (OA&M) functions, and the architectural requirements of such functions. The Switching and Signaling networks are also examined, as is the future of Operations Architecture.

Chapter 4, "Network Components" looks at the physical elements that comprise the Intelligent Network. There is a general overview of network topology and a detailed examination of the main components: the Service Control Point (SCP), the Signaling Transfer Point (STP), the Service Switching Point (SSP), and the service user and service subscriber interfaces. These elements are described in

---

[1] This report uses "CCS7" to refer to both the Signaling System Number Seven (the term used in the United States) and the Common Channel Signaling Number Seven, which is the CCITT term.

terms of the hardware and software that is currently proposed to implement the network and services.

This chapter also considers the future of the IN elements: how IN/1 components can be designed for the transition to IN/2, and areas for further standardization.

Chapter 1. The Intelligent Network

1.1 IN Architecture and Capabilities

The Intelligent Network (IN) is a telecommunications network services control architecture.

The goal of this services control architecture is to provide a framework so that the Network Operator can introduce, control and manage services more effectively, economically and rapidly than the current network architecture allows.

The main benefit of the IN architecture is the possibility to improve the quantity, and to develop new sources, of revenue. This is particularly desirable in an environment with a high penetration of available services per capita. For most environments, the IN will be a stimulation or basis for revenue generation, in both the short and long term.

In a competitive environment the IN will maintain and enhance existing revenues by providing IN-based services which offer competitive alternatives to Bypass and Private Networks. The USA is currently experiencing a level of competition unmatched elsewhere in the world. Similarly, many other countries have or are preparing to enact deregulatory policies which will create and expose their PTTs to competition. Thus, the importance of IN architecture (and competitive countermeasures, in general) is increasing worldwide.

The technical features that the IN architecture must provide are:

- Network connection control intelligence at centralized nodes. This node is known as the Service Control Point (SCP).
- Network nodes which switch connections under the direction of the SCP. These nodes are Service Switching Points (SSPs).
- Standard Network interfaces (for example, CCS7 and ISDN) at points such as the SCP and SSP. These interfaces facilitate competition between suppliers of network and service products, and thus stimulate multiple vendor environments.
- Rapid and economic **service creation** capabilities (such as customer programmability and program portability), for example, a function by which the network operator can identify a need, create or acquire a corresponding service, and deploy it within a market opportunity window, while maintaining the integrity of the network.

1.1.1 IN Goals

In February 1985, a Regional Bell Operating Company (RBOC) submitted a
Request for Information (RFI) for a Feature Node concept (see Feature
Node/Service Interface Concept, SR-NPL-000108) with the following objectives:
- Support the rapid introduction of new services in the network.
- Help establish equipment and interface standards to give the RBOCs the widest
  possible choice of vendor products.
- Create opportunities for non-RBOC service vendors to offer services that
  stimulate network usage.

This Feature Node concept was later adopted by Bellcore, and enhanced to become
the **Intelligent Network concept**.

At the direction of Bell Atlantic and the other RBOCs, Bellcore has since
published a number of documents and held seminars describing the IN concept.
Their IN objectives have been stated as providing:
- Flexible network architecture
  - Adaptable to rapidly changing technical, regulatory, and marketing
    environments
  - Independent of specific services or capabilities
  - Greater customer (PTT) control of service features and functions
  - Efficient network control and administration
  - Telecommunication infrastructure supportive of national needs.
- Standard network interfaces
  - Promoting a competitive environment
  - Consistent with regulations
  - Stimulating use of the network
  - Standards
    - Signaling network (CCS7 and Transaction Capability Application
      Part—TCAP)
    - Application messages (Functional Components)
    - Standard capabilities and procedures (for example, ISDN).
- Rapid service introduction
  - PTT programmable
  - Ability to meet market window
  - Services independent of network transport mechanism (for example, the
    same Closed User Group structures for Packet Switching, Centrex, ISDN)
  - Potential for ubiquitous services
  - Multiple suppliers
  - Ability to service niche markets and individual customers.

The following example illustrates the RBOC's interest in those objectives:

> The 800 service in the USA generates several billion dollars in
> revenue per year, the number of subscribers grows by 10% per
> annum, and revenue increases by 20% per annum.

Figure 1 shows that typical RBOC's annual revenue is around $10 billion. Improving their network utilization by 1% would generate $100 million extra revenue.

	Access Lines (millions)	Revenue ($ billions)
Regional Bell Companies		
Ameritech	14.6	9.0
Bell Atlantic	15.1	9.1
BellSouth	14.5	10.7
NYNEX	13.6	10.3
Pacific Telesis	11.7	8.5
Southwestern Bell	11.0	7.9
US West	11.2	7.8
Other U.S. LECs		
GTE	10.2	15.7
U S Telecom	3.3	3.2
Contel	2.2	2.6
SNET	1.7	1.3
Alltel	1.3	0.7
Centel	1.3	
Cincinnati Bell	0.6	0.5
Rochester Tel	0.5	0.4
Others	0.6	
Independent U.S. Companies		
AT&T		36
PBX operations	6-10	0.7
IBM		50
Rolm (PBX)	3-6	0.4
Wang	2.6	
Intercom (PBX)	0.5 - 2	0.1

Figure 1. Access Lines and Corporate Revenue, 1985[2]

---

[2] From the Geodesic Network, 1987 Report on Competition in the Telephone Industry, Peter. W. Huber.

1.1.2 IN Technical Overview

IN Components

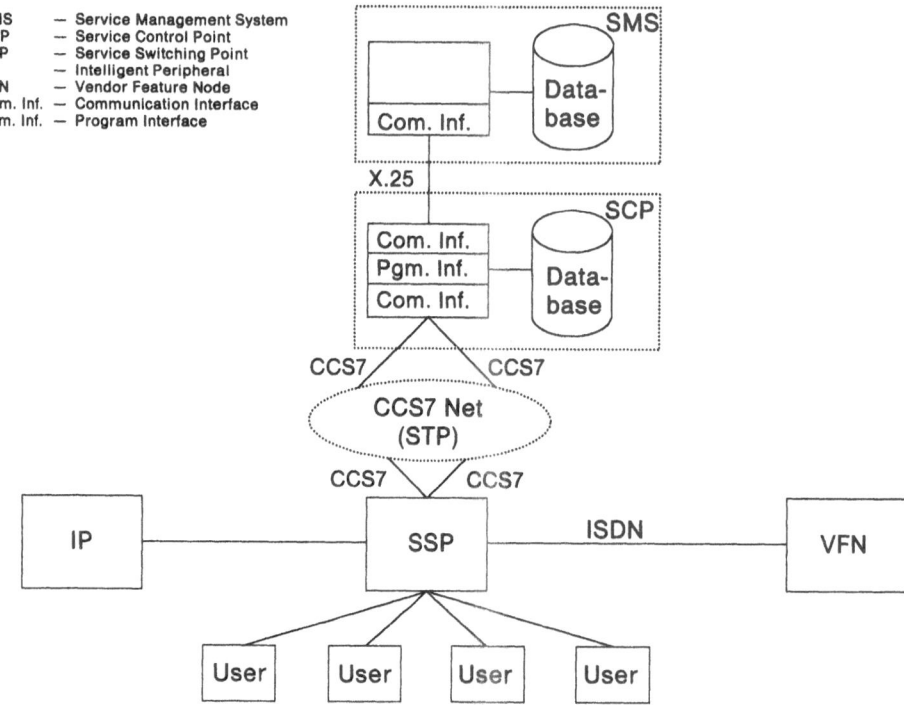

SMS	—	Service Management System
SCP	—	Service Control Point
SSP	—	Service Switching Point
IP	—	Intelligent Peripheral
VFN	—	Vendor Feature Node
Com. Inf.	—	Communication Interface
Pgm. Inf.	—	Program Interface

Figure 2. IN Components

Figure 2 shows an overview of the joint study vision of the IN.

The IN's main advantage is the ability to orchestrate exchange service execution from a small set of Intelligent Network nodes known as Service Control Points (SCPs). SCPs are connected to the network exchanges (known as Service Switching Points (SSPs)) via a standardized interface CCS7. The CCS7 will facilitate a multi-vendor SCP and SSP marketplace, and the standardization of application interfaces allows a multi-vendor software marketplace for SCP applications (that is, the service control logic and its related data).

The Service Switching Points detect when the SCP should handle a service. The SSP forwards a standardized CCS7 (TCAP) message containing relevant service information.

Via the TCAP message, the service control logic in the SCP directs the SSPs to perform the individual functions that collectively constitute the service (such as connecting to a subscriber number or an announcement machine).

The IN's long term goal is the ability to introduce new services, or change existing services quickly, without having to adapt SSP software (only parameters or trigger updates). The adaptation will be confined to the SCP where parameters or stimuli are updated. This goal will be achieved in stages. This report is primarily concerned with the IN's initial form, known as IN/1. This does not require that services can be introduced without affecting SSP software. However, ideas and possibilities associated with the IN progression are discussed as well.

Stage 1: IN/1

IN/1 requires updates in the SSP and SCP in order to support a new service. A typical IN/1 service is the Green Number Service (GNS) with which a subscriber can call a number free of charge.

The SSP contains triggers (such as the value of the dialed digits) that tell the SSP to send a message to an SCP in order to get information about the destination to which the call should be routed. Migration from IN/1 to IN/2 implies significant changes in the SSPs to accommodate new services.

Stage 2: IN/2

Once IN/2 is in place, no updates need be made to the SSP's software when new services are introduced. The IN/2 triggers advise the SSP whether to let an SCP handle a service request or whether to complete execution locally. All SSPs and SCPs contain a set of basic service elements (for example, connect two lines, disconnect a line). The SCP also contains service relevant data. These basic service elements are known as functional components (FCs) from which each service can be constructed. A customer could conceptualize a new service and the network operator, via the SMS/SCP, could construct it quite rapidly. Any successful and widely-used service may be downloaded (via the service logic) to, but transparent to, the SSPs (if this is more economic or provides a desired higher grade of service). This facilitates complete Rapid Service Creation. Rapid Service Creation and User Programmability will take place in the SCP and the SMS. There will probably be one or more interim stages between IN/1 and IN/2, for example IN/1 + where the SSP provides increasing flexibility in accommodating rapid service creation.

1.2 IN Definition

1.2.1 IN Elements

Figure 3 gives an overview of the IN elements. These are used by:

Service User (A): Who takes advantage of the service by, for example, dialing a set of digits.

Service Subscriber (B): Who, via the network operator's service access, provides the user's end service.

Network Operator (C)[3]: Who provides the control logic and network (service) enabling the service user and subscriber to do business.
Note: The service subscriber and network operator can be the same entity.

Network Product Supplier: Who supplies products enabling the network operator to provide service control.
The following is a definition of the capabilities required from each IN element:

* **Service Management System (SMS)**
  The SMS is owned by the network operator. It updates SCPs with new data or programs and collects statistics from the SCPs. The SMS also enables the service subscriber to control his own service parameters via a terminal linked to the SMS. (For example, the subscriber may define the day and time when an "800" number should be routed to a specific office.) This modification is filtered or validated by the network operator. The SMS is normally a commercial computer such as the IBM/370 or Siemens 7.5XX (see Chapter 3,

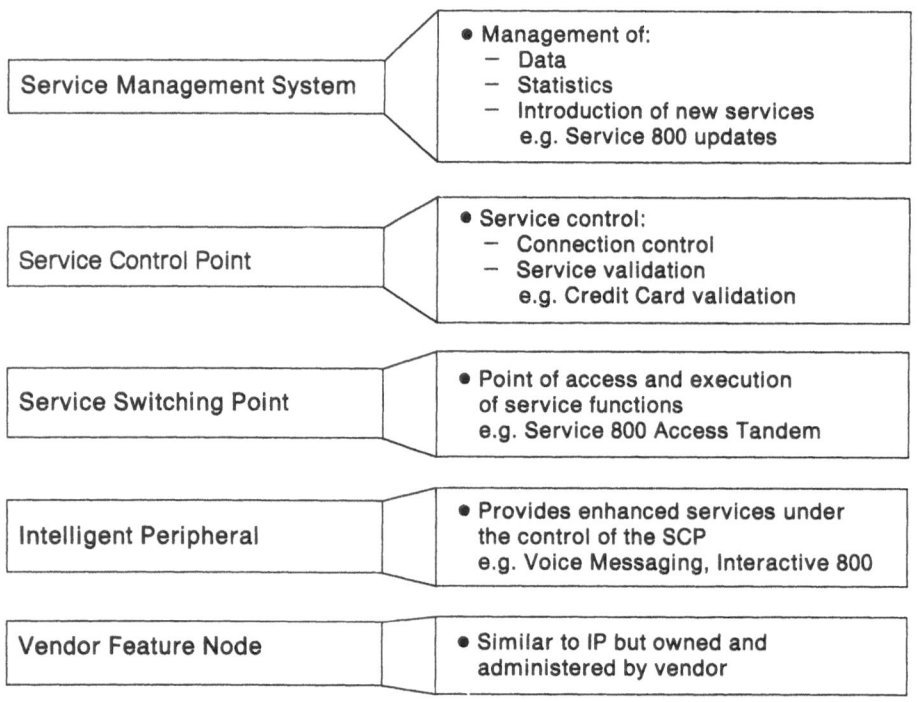

Figure 3. IN Elements

"Network and Service Management"). The SMS may also provide a
development environment for new services.

- **Service Control Point (SCP)**
The SCP is used when new IN services are introduced into the network and
then activated. If a service is based on functional components (IN/1 + and
IN/2), the FCs are executed with the help of a Service Logic Interpreter (using
an explanatory script). Some SCP services may require large amounts of data,
which must reside on direct access storage devices, for example, disks. The
service programs and the data are updated from the SMS. The SCP is a
commercial computer or a modified switch. The critical factor is that the unit
can access databases efficiently and reliably and provide a software platform for
rapid service creation (through "user programmability" and "program
portability").

- **Signaling Transfer Point (STP)**
The STP is part of the Common Channel Signaling Number Seven (CCS7)
network. It switches CCS7 messages to different CCS7 nodes. The CCS7 is a
standardized communication interface through which the goal of multivendor
SCPs and SSPs can be achieved. The use of a standalone or integrated STP
will depend on network specific configurations. The STP is normally produced
by traditional switch manufacturers.

- **Service Switching Point (SSP)**
The SSP serves as an **access point** for service users and executes heavily used
services (as described for the SCP). In IN/1, no "user programmability" exists
in the SSP. The SSP is produced by traditional switch manufacturers. The
TCAP will be used to communicate between SSPs and SCPs.

- **Intelligent Peripheral (IP)**
The IP provides enhanced services, controlled by an SCP or SSP. It is more
economical for several users to share an IP because the capabilities in the IP are
either not in the SSP or are too expensive to put in all SSPs. The following are
examples of typical IP functions:
  - Announcements
  - Speech synthesizing
  - Voice messaging
  - Speech recognition
  - Database information made accessible to the end user.
The IP is usually accessed from the SSP via a circuit or packet basis, such as
ISDN.

- **Vendor Feature Node (VFN) or Service Provider (SP)**
The VFN is outside the network; it is owned and administered by a service
subscriber. It can provide many services. The VFN provides the services
mentioned for IPs and may be connected via a CCS7 link. However, this type
of connection includes filter logic which prevents a VFN using CCS7 messages
which could interfere with network operations.

1.2.2 An IN Service Example

Figure 4 shows how the non-IN Green Number, or "freephone" Service operates. A service user dials a number beginning with 800, for example, 800-NXX-7800. The local exchange does normal translation and routing to a switch higher in the network. This switch in turn connects to even higher level exchanges which contain the 800 database.

The higher level exchange translates and routes connections through the network to the subscriber being called. As shown in Figure 4, this method is resource-intensive because of the trunks held and the CPU processing required. Thus in large networks, several higher level 800 exchanges are required to support this service and improve its efficiency. This in turn makes updating and administration of the service more complex.

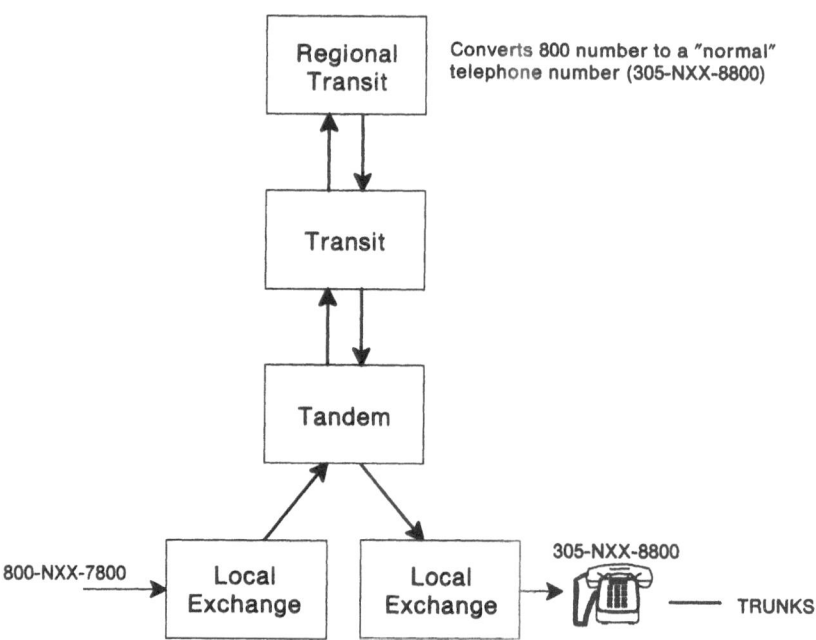

Figure 4. Non-IN GNS Service

Figure 5 shows how the IN/1 Green Number Service operates.

A service user dials a number, for example 800-NXX-7800. While translating the number, the local exchange detects a trigger in the SSP database telling it that this 800 number is a pseudo-number which must be translated by an SCP. The local exchange sends a TCAP message (containing the 800 number dialed and other information) over the CCS7 network to an SCP. The SCP uses the 800 number to access a database containing the 800 number's corresponding directory number. (This number could depend on factors like day, time of day, origination, and so

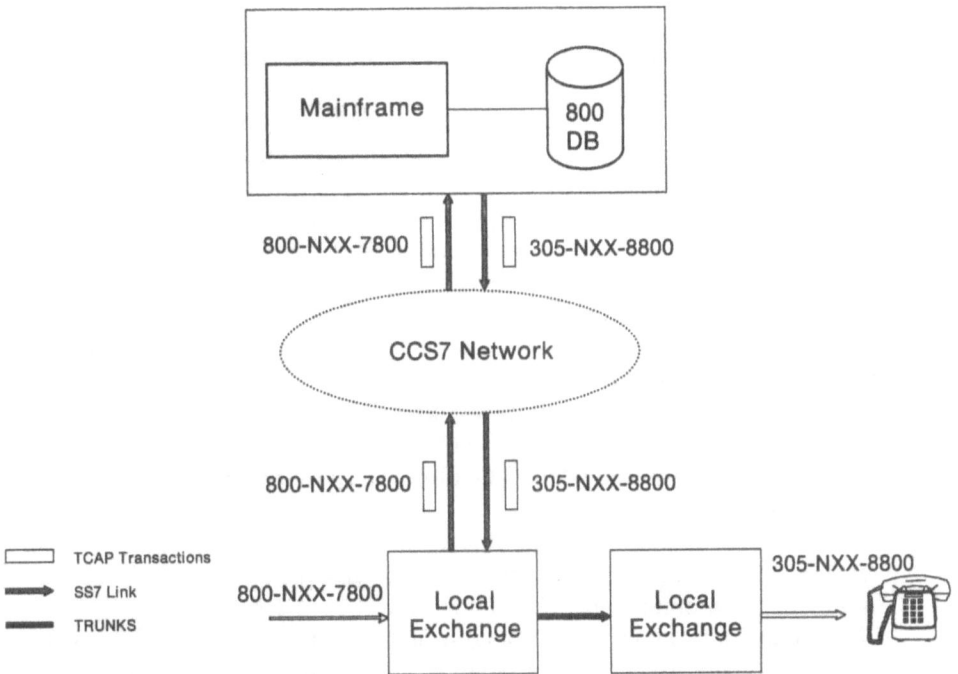

Figure 5. IN-Based Green Number Service

on.) The SCP sends the directory number (in this example 305-NXX-8800) to the local exchange in another TCAP message. The local exchange uses the directory number to execute translation and routing to the subscriber called.
Figure 5 shows that this solution uses network resources more efficiently. If new 800 numbers are added, updating need only be done in the SCP database.

This principles of trigger detection and database dialog are the basis of all proposed IN services. These services are discussed in later parts of this report and include:

- Green Number Service (GNS)
- Alternate Billing Service (ABS)
- Emergency Response Service (ERS)
- Private Virtual Network (PVN)
- Area Wide Centrex (AWC)
- Pay per View (PPV).

The basic principles of trigger detection and database dialog remain applicable, even as more complex service dialogs (involving an IP or VFN) are required.

1.3 IN-User Programmability

The function known as "customer (network operator) programmability", includes "application portability" and is a response to the IN business goal of giving the network operator "rapid service creation": rapid enough to allow market opportunities to be seized and to increase the overall speed of service creation. It also enables network operators to expand their sources of software services beyond traditional telecommunications market suppliers. Therefore, the network operator's goal is not necessarily the simple aim of being able to write software service programs himself, but to provide an environment (usually a standard platform in the SCP) that allows many vendors to write and market software services that can run on different hardware systems. This will result in a more competitive environment that will accelerate service creation. Thus, the SMS must provide an environment for quick and simple service creation and testing, using standard program interfaces to allow multi-vendor participation. The SCP and SSP must also provide a set of capabilities which will allow the following network evolution:

- IN/1
    - Standardized software platform in the SCP enabling the simple addition of a service to the SCP, for example, database manager, operating system (OS), TCAP monitor, and application service program
    - Real time control (such as overload protection) and efficiency in the SCP
    - An SMS facility to administer services and provide parameter driven changes to service logic
    - Standard software language(s) allowing an effective multi-vendor environment for the SCP and SMS
    - Standardized interfaces between the SMS, SCP, SSP, IP, and VFN.
- IN/1 +
    - Application Program Interface providing the first version of PTT, or user programmability
    - Ease of service dialog creation (scripting) in the SSP via trigger tables when a new service is introduced in the SCP
    - Service portability from the SCP to the SSP
    - Functional components (first set).
- IN/2
    - Functional components (full set)
    - An SMS service creation environment to develop, create, and test services which will later run in real time in the SCP or SSP.

Summary
It must be possible to adjust parameters (for example, time, location, and carrier) quickly within a service. It is also necessary that individual service providers (for example, "800 number" subscribers) be able to activate this function. The achievement of rapid service creation will occur in stages. The key goal remains the ability to create services more rapidly in a multi-vendor software services environment.

1.4 IN Introduction Scenarios

Various national and regional telecommunications networks currently have very different technology configurations, evolution plans, service user demographics, and subscriber topography.

An examination of some technical variables and fundamental evolution plans will help create some introduction scenarios and causal rules which (based on an existing network configuration, the service user's requirements, and a defined set of target IN services) will produce an introduction strategy.

An introductory strategy for IN services aims to maximize the following sequential objectives:

1. Service user benefit
2. Service availability to service users
3. Service revenue generation to the network operator
4. Service profit to the network operator
5. Network operator's competitive position
6. Use of network product suppliers' standard products (hardware and software) in order to minimize service-related development cost.

A model of service-supporting network capabilities is used to evaluate what is needed to achieve these or any objectives.

1.4.1 The Model

This model assumes that the introduction methodology has the following dimensions:

- Service categories
  The network introduction strategy depends on the service characteristics.
- Network base
  The introductory strategy depends on existing network characteristics, such as the CCS7 penetration at particular network levels.
- Service User Access (signaling)
  The strategy will vary depending on the penetration of analog, analog multi-frequency, and ISDN terminals in the network. The equipment available behind the access customer premises equipment may also affect the strategy, but is not considered here.

Service Categories
The services being implemented or considered for the IN can be categorized as follows:

- A **B-Number (called number) Service** based on the SCP receiving knowledge of the number dialed. It is not necessary to know the A party (originating party) number; for example, the Green Number Service relies only on the B-number and global origin (country or region of country) for service execution. These services can be introduced in all network configurations on a broad scale more easily than the other services. The Green Number Service (without A number) is a B-number service.

- The **(A + B)-Number Services** rely on the SCP receiving knowledge of the number dialed and from which directory number the call originated. These services require the CCS7 to penetrate the network more deeply. Alternatively, the services require other supportive signaling capabilities (for example ANI) which deliver the A-number from the local originating exchange to an SSP from which it is communicated to the SCP.

  The following are (A + B)-number services:
  - Enhanced GNS (with A-number)
  - Emergency Response Service
  - PVN
  - Area Wide Centrex
  - Alternate Billing Service.

- The **Interactive Services** rely on the SCP receiving knowledge of the A and B party number and decision data (which is dialog prompted). The decision data searched for includes calling card number, passwords, and messages. Calling card and interactive GNS services use this SCP function. These services require:
  - IP (or VFN) availability for executing the dialog prompting with the A party and communication of results to the SCP.
  - A-party signaling capability (such as MF or ISDN) to facilitate the dialog with the IP.
  - The (A + B) number services function of forwarding the A party number to the SCP.

Interactive services are the most demanding on network configuration capabilities. The following are Interactive Services:

- PVN
- Call Completion
- Voice Messaging
- Anywhere Call Pick up
- Private transaction network
- Bank at home
- Pay per view
- Calling card verification
- Videotex Interactive.

Network Base

Three network configurations are considered in the model:

C1. Conventional, that is, no CCS7 penetration. This is a network composed of conventional analog and digital transmission and switching, with little or no CCS7. It is a common configuration around the world.

C2. Tandem and above CCS7 penetration. This network is based on C1 but has developed by introducing CCS7 into its long distance and regional transit nodes.

C3. Large scale tandem and local CCS7 penetration. This is an extension of C2 which has pushed the penetration of CCS7 deeper and broader into its local exchanges with additional capability for ISDN.

Therefore, C1, C2, and C3 are stages in the progression from conventional in-band signaling to ISDN.

Network Services User Access

Two types of service user access are considered in the model:

1. Non-ISDN plain old telephone service (POTS) access, such as dial pulsing[4] and MF.
2. ISDN basic access (residential and business) and primary access business.

Conventional instruments are most widely used, since ISDN is only now being introduced. However, assuming the exchange is not an ISDN island, ISDN is equivalent to having CCS7 at the local exchange. It is particularly suited for IN dialog and is therefore included in the model. ISDN is targeted at the business customer base, and this segment may require a specific set of IN features. IN is a service control architecture while ISDN is an access control architecture, but they are complementary and enhance each other. They are not, however, dependently linked.

1.4.2 Introduction Scenarios

Table 1 shows the model's three dimensions. Each box is a coordinate containing an introductory strategy, based on the status of CCS7 network penetration, the service category to be introduced, and the service user access. Table 1 shows that for conventional instruments, the introduction strategy is simplest for B-number services. Only in the C1 network is the base (tandem access CCS7) unavailable and required before the SCP can be brought online. For interactive services, however, every network configuration must be enhanced before the SCP can be brought online. Generally, the introduction effort for IN services for conventional instruments reduces as CCS7 penetration increases.

To provide IN services to users with ISDN access requires less network enhancement, regardless of which service category is considered. This is because ISDN implies the existence of CCS7 at the local exchange (unless it is an ISDN island) and the existence of a dialog capability for signal prompting. Therefore, networks that tend toward ISDN are particularly receptive to IN services in general, especially more complex IN features.

The model deals primarily with a method for determining what network configuration steps and alternatives are associated with introducing IN services. It is necessary to identify some of the logic necessary to select alternatives, that is, does the cost/benefit ratio justify its introduction?

---

[4] Dial pulsing instruments only allow meaningful usage of the interactive services when real intelligence exists in the network, for example speech recognition.

Table 1. IN Introduction Evolution Scenario

Access Type	Service Category	C1 Conventional (no CCS)	C2 TANDEM CCS	C3 TANDEM + LOCAL CCS
POTS	B number	• Introduce C2	Basis Exists	Basis Exists
	(A + B) number	• Introduce C2 • Introduce A forwarding or C3	• Introduce A forwarding or C3	Basis Exists
	Interactive	• Introduce C2 • Introduce A forwarding or C3 • Introduce IP/ VFN	• Introduce A forwarding or C3 • Introduce IP/ VFN	• Introduce IP/ VFN
ISDN	B number	• Introduce C2*	Basis Exists	Basis Exists
	(A + B) number	• Introduce C2*	Basis Exists	Basis Exists
	Interactive	• Introduce C2* • Introduce IP/ VFN	• Introduce IP/ VFN	• Introduce IP/ VFN
* = If exchange is an ISDN Island				

A network operator may benefit from a service by:
- Increased network operator traffic revenue due to usage or increased usage of an IN-based service by a specific set of service users
- Decreased costs or overheads for providing an IN-based service.
- Increased service provider revenue (when the network operator is the service provider) due to usage or increased usage of an IN-based service by a specific set of service users.

The cost of upgrading the target network associated with a service relates directly to the service user base, geographical distribution, the service user access available, and the current network configuration. Use Table 1 to determine the cost categories.

Examples
1. An existing C3 network involves the least cost to the network operator in introducing any type of service to service users.
2. With C2, introducing (A + B)-number and interactive services to POTS users (maximize service availability) will cost more than making them available to ISDN service users only.

Table 1 indicates the factors to be considered regardless of which objective is chosen. The major business decision to be made is: whether to provide a service to the POTS market segment, requiring a large investment for a large market, or whether to provide a service to ISDN users, which requires a smaller investment but which is for a smaller market base.

1.5 IN Architecture Validation

The IN architecture can be validated by applying a complex service as identified by AT&T[5] which will put stress on the architecture.

The **Pay Per View** service is chosen because it is complex and represents the type of service likely to become increasingly popular.

The Pay-Per-View service is a television entertainment service whereby consumers watch a scheduled movie or live event on their home television set and are charged for the specific viewing time. It is a "usage-sensitive" pricing system for television entertainment, similar to the movie services found in some hotels. The viability of Pay-Per-View depends on an efficient system for ordering, distributing and billing. Its operation is shown in Figure 6.

In a typical Pay-Per-View system, the consumer orders a specific program by placing a telephone call to the local cable company. An attendant enters the information into the company's business management computer system. The program is transmitted scrambled; descrambling occurs only in the homes of those who ordered the program. The consumer is charged according to the cable company's own billing system.

This method has several major limitations. It is expensive, because consumers tend to wait until a few minutes before the program begins before they place their orders. The resulting high volume of orders requires many human attendants and service lines. Regardless of the number of attendants, many calls are lost and consumers are dissatisfied. The costs of manual order entry take up a lot of the program revenue, with labor costs particularly high during typical consumer viewing periods (evening, night, and weekends). Cable companies try to discourage last-minute calling by offering incentives for ordering early; however, this requires that consumers plan their viewing in advance and substantially reduces the number of orders made, and thus reduces revenue.

In addition, attendants are often overwhelmed by the volume of calls (regardless of the number of attendants) so that calls are lost and customers are dissatisfied.

Some cable companies have tried automated voice response systems and other devices to solve these problems, but with limited success. A local approach means that order-taking effectiveness varies widely and rules out nationwide marketing techniques. The Pay-Per-View industry needs standard and consistent order-taking technology.

---

[5] AT&T Technical Journal, May/June 1987, Vol 66, Issue 3.

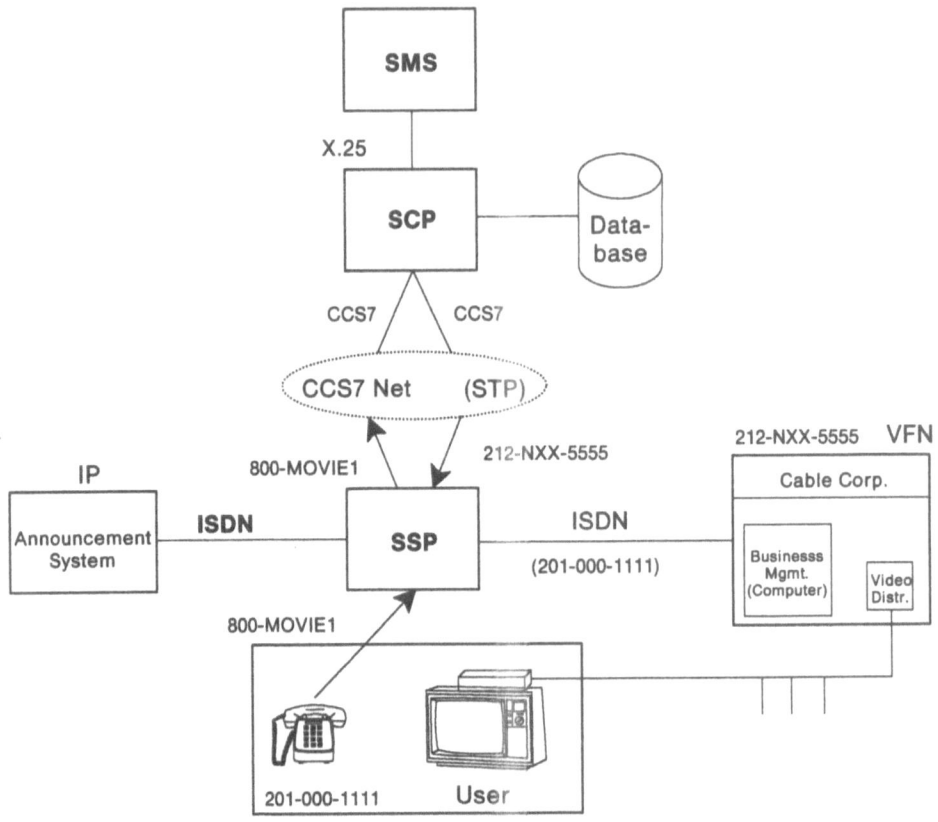

Figure 6. IN-Pay-Per-View

An IN architecture for Pay-Per-View must have following characteristics:
- Ability to handle a large volume of last-minute orders
- Ease of use
- Ability to work with both dial-pulse and dual-tone multifrequency (DTMF) phones
- Minimum initial investment, that is, supported by an architecture that requires no additional hardware in the consumer's home
- Full automation from end to end, including an automated interface with cable companies business management and billing systems.

The required IN architecture functions are:

A GNS base

Which allows consumers to place toll-free calls. Each movie or event is assigned a unique GNS number which the consumer dials to order the movie or event.

Announcement system (IP)

Within the network a high traffic capacity to provide quick confirmation of orders.

Automatic Number Identification (ANI)
> To retain caller's telephone number. The ANI information is collected and delivered to a Pay-Per-View distributor. Distributing the information to the cable companies can be handled using ISDN.

An Order Processing System (VFN)
> At the cable company, to receive ordering information and use it to activate the addressable system and the cable company's billing process.

The cable companies also need a connection to the SMS to update the movie GNS numbers.

This service requires the use of all IN components and particularly stresses the real time aspect, in that service users concentrate their utilization. This requires engineering and monitoring of the real time availability of all components, especially the IP and VFN.

The service reaffirms the necessity for A-number forwarding (ANI) when the service user is not connected to a local exchange equipped with CCS7. Additionally, a dialed GNS number is used as a substitute for interactive selection, that is, to send a code for the desired movie. This makes the service available to service users with POTS access at the cost of Green Number Service numbers.

The architecture can handle this type of service. Furthermore, it provides a set of options (local exchange CCS versus ANI, and GNS numbers versus interactive dialing) which accommodate different network environments. What remains to be analyzed is the ability to handle a large number of transactions compressed into very active time periods (bursts of service user activity). However, since the critical network elements will probably be the VFN and IP, the architecture enables the deployment of multiple element groups reachable via primary access(es) ISDN.

Chapter 2. Functional Characteristics Common to Selected IN Services

2.1 Overview

This chapter is a high level functional description of characteristics common to five services: GNS, ABS, ERS, PVN, and AWC. These services were selected by Bell Atlantic, IBM and Siemens on the basis of potential revenue (for the network operator).

This chapter examines the standards to be applied to each service, and how the services will be used. The functional characteristics described include billing, service logic, databases, traffic measurement, and performance requirements.

2.2 Methodology

The service descriptions are primarily based on Bellcore information: Technical Requirements (TRs), Special Reports (SRs), and Technical Advisories (TAs). Services have been examined according to an IN/1 architecture consisting of stored program controlled exchanges. These exchanges are connected by trunks with existing signaling. This signaling may include:

- Dial pulse signaling
- Dual-tone multifrequency (DTMF)
- CCS6
- CCS7.

The IN/1 architecture consists of one or more SCPs serving as a centralized database system and connected to the SSPs via:

- CCS7 links if accessed from the PSTN
- X.25 links if accessed from the Packet Switched Public Data Network (PSPDN).

The SCP is connected to the SMS via BX.25[6] links.

The services are described pursuant to what is known of service implementation in the United States, however, adaptations to the requirements of individual countries are also considered. **It must be emphasized** that the service descriptions are hypothetical implementations only; actual implementation will address the needs of the particular country's telecommunications network.

The network elements examined in Chapter Four must satisfy the requirements specified in this chapter.

2.3 Summary of Findings

- In a network with several vendors, it is essential that there are standards to ensure the proper introduction and maintenance of services.
- Service functions are largely independent of network signaling but require that call-origin information be passed through the network.
- It is necessary that the SCP database can be split or replicated and distributed to more than one SCP for security and load sharing reasons. Replicated databases can be stored in the same SCP or distributed.
- Billing and traffic measurement practices vary considerably between countries. The PTT's method of billing, for example, will influence the SCP application program development and the transaction rates between the SCP and the SSP. Country-specific billing methods will significantly affect the engineering of the SCP.
- It is expected that the service subscriber will require more control over telecommunication services in the future. With GNS, ERS, PVN, and AWC, the service subscriber can control his own service parameters, either directly (via a terminal and a dial-up connection) or via a service request.
- The interaction of IN services with other services, both IN and non-IN requires further investigation. For example, if a Green Number Service user in the United States receives a busy tone and he were able to activate the CLASS automatic recall, the impact of this interaction on network functions and dynamics could be significant. This type of scenario must be explored.

---

[6] The BX.25 is the Bellcore X.25. This is the protocol used in the USA to communicate with operation support systems. In Europe, an OSI model using ISDN or X.25 is being considered.

2.4 Standards

Telco and service standards must be based on country-specific requirements for service functions.

The CCS7 (TCAP, SCCP, and MTP) must be based on CCITT standards or Bellcore requirements, depending on the country of deployment.

Service specific TCAP messages are listed under the individual service description.

The following X.25 requirements are for the US only and may be different from country to country. The signaling protocol between:

- The SMS and the SCP is based on X.25 according to CCITT
- The SCP and the SMS is based on the BX.25 protocol
- The SCP and Operating Support Centers is based on the BX.25 protocol
- The PSPDN and the SCP is based on X.25, according to CCITT.

The X.25 and BX.25 protocols are very similar on the lower three levels of the Open Systems Interconnection (OSI) model.

Trunk signaling between the network exchanges may vary from country to country. This is considered under service specific descriptions.

ISDN access will be based on country-specific requirements. "Appendix A. Supplementary Services" gives a list of ISDN supplementary services as currently known from the standardization bodies CCITT and CEPT.

2.5 Service Interaction

2.5.1 Service User

It is assumed that the service user instrument is a telephone set or ISDN Customer Premises Equipment (CPE) and that the service is fully functional for both types of CPE access.

This excludes any prompting between a service and a telephone user with no multifrequency (MF) dialing function, as dialog-interaction is extremely difficult, or impossible with dial pulse signaling.

Guidance: Service specific.

Data: The data a service user provides depends on the individual service.

2.5.2 Service Subscriber

The service subscriber can change service parameters in three ways:

- By submitting a written service request to the network operator, who is responsible (via a service administration center) for updating the subscriber's records.
- By calling a network operator service center and verbally notifying an operator of the changes desired. The network operator personnel update the subscriber's records (via a service administration center).
- Via a data terminal (not available for ABS). The subscriber must be able to dial into, and access the SMS. A log-on process provides security mechanisms to control proper access.

A subscriber can request billing and traffic information, which can be generated in a list format, or as summary tables. This information can help a subscriber optimize his business. For example, with GNS, a list of callers, their numbers and data, time when the call was made, and how often the subscriber's number was busy provides valuable demographic information.

A special option allows the subscriber to make express updates, ensuring that the desired change is operational within 15 minutes.

Guidance: The service subscriber defines his customized service features using menus and structured support (for example, a routing tree). Subscribers can modify existing service information, but must apply to the Telco if they want a new service. Each data change initiated by a service subscriber is confirmed on the screen, and the subscriber can get a log of the updates made to his service parameters.

Data: The data a service subscriber receives and supplies depends on the individual service.

2.5.3 SMS Access Instrument

There are two ways of connecting terminals to the SMS[7]:

Direct Access: The directly connected service subscribers access the SMS via synchronous 3270 terminals using 9.6KB dedicated circuits. Two circuits should be used for each terminal site. The second circuit can be used as a backup circuit or to operate a printer.

Dial-In Access: Using various hardware components.
Terminals must support:

- 24 lines with 80 characters per line
- Standard EBCDIC character set
- At least 12 program function keys

---

[7] SMS/800 Terminal Data Communications Planning Information, Bellcore Special Report, SR-STS-000742.

- Reverse video or highlighting
- Audible alarm
- PA1, PA2 and TAB keys.

Note: Future terminals should also be able to access the SMS via an ISDN interface.

2.5.4 Network Operator

The network operator administers and maintains service-specific data and logic. The activities of network operator personnel include[8]:

- Maintaining reference material including service specific security tables
- Scheduling the SMS activities, such as transmission of updates, report generation, database backups, and so on
- Modifying database records because of changes in user related data
- Adding new service subscribers' data to the database
- Updating database records
- Performing database audits and integrity checks to resolve database inconsistencies
- Performing database backup and recovery
- Making mass changes to the service database
- Defining parameters for traffic measurements and billing
- Defining parameters for service reports
- Administering numbers
- Administering database
- Generating scheduled reports with a preselected format and contents
- Preparing ad hoc reports for the network operator
- Providing special reports for service subscribers
- Splitting of the database into more SCPs
- Managing traffic flow between SCPs.

Guidance: In addition to the guidance that the network operator receives when performing the above-mentioned functions, the network operator is guided when:

- Defining security mechanisms, administering logons and passwords, assigning database access and functional restrictions.
- Accessing service documentation and archived data online.

Data: The data related to the network operator is highly proprietary and service-specific.

---

[8] Plan for the Second Generation of the Intelligent Network, Bellcore Special Report, SR-NPL-000444.

2.6 Billing

Billing requirements are country-specific. The critical point is that billing data must be collected and stored on a per call basis. In general, the two methods of charging are automatic pulse metering and toll ticketing.

2.6.1 SSP

Billing may be done in the originating exchange, and billing records are later transferred, for example via an X.25, to an administrative center for further processing.

2.6.2 SCP

Billing data may be collected in the SSP or SCP. The SSP must be able to record and deliver billing data to the SCP. The SCP must record and supply service features information to the SSP billing.

After billing data is collected at the SSP or SCP, it is formatted into billing records which are automatically transferred to the SMS (for billing statistics) and to the system that generates bills (possibly the SMS).

2.6.3 SMS

There are three kinds of billing data that can be collected in the SMS:
- Billing for subscription to a service; since the subscribing is done via SMS, billing should also be performed via this system.
- Billing for SMS access when a subscriber modifies his service parameters, makes queries or requests reports on billing or traffic statistics. Based on the use of computer resources, the SMS builds and stores the corresponding billing records.
- Billing for use of service features. This should be done at the SMS because all necessary data is available there.

The service subscriber pays a monthly fee for service subscription and also pays for being able to update his records.

2.7 Service Logic

The SCP comprises the SCP node, the SCP platform, and applications. The node performs functions common to applications, or independent of any application; it provides all functions for handling service-related, administrative, and network messages. These functions include message discrimination, distribution, routing, and network management and testing. For example, when the SCP node receives a service-related message, it distributes the incoming message to the proper application. In turn, the application issues a response message to the node, which routes it to the appropriate network elements.

The SCP node gathers data on all incoming and outgoing messages to assist in network administration and cost allocation. This data is collected at the node, and transmitted to an administrative system for processing.

The SCP node also measures the frequency of SCP hardware and software failures, network link failures, resource usage, overload counts, and so on. It provides information needed to perform maintenance procedures, thus minimizing the impact of failures on system performance. The node may take action to prevent and correct overload at the node or at a particular application.

Applications: An application served via the SCP provides a variety of services over the CCS7 network and/or X.25 network facilities. The services in a single application generally have a common base of customer-specific data. The applications contain all information necessary to respond to incoming queries. Responses may provide verification of numbers for billing purposes or customer line information, or provide instructions essential for call completion. Applications operate on an SCP software platform that offers services like database access and administrative support.

Messages Sent Between IN Elements
A TCAP message is comprised of **components** specifying an action to be taken, reporting the result of a requested action, or reporting an error condition. Table 2 lists the TCAP messages sent between the SSP and the SCP. The application-specific content of these messages varies. Service specific TCAP messages are described in the individual service descriptions in parts 2 to 6 of this report. TCAP messages include the following component types:
Invoke
> Requests that an operation should be performed.

Return Result (Not Last)
> Where the TCAP application program segments the result of an operation, the result is sent in components of this type. The last segment is sent in a Return Result (Last) component.

Return Result (Last)
> Reports the successful completion of an operation. It contains the last, or only segment, of a result in the transaction user information.

Table 2. TCAP Messages Between SSP and SCP									
	←	Operation Code → of Component							
TCAP Compo- nent Type	Family Name	Specifier	SSP to SCP	SCP to SSP	GNS	ABS	ERS	PVN	AWC
Invoke	Provide Instructions	Start	X		X		X	X	X
Invoke	Connection Cntrl.	Connect		X	X		X	X	X
Invoke	Network Mgmt.	Automatic Call Gap		X	X	X	X	X	X
Invoke	Send Notification	Termination		X	X			X	X
Invoke	Caller Interaction	Play Announc.		X	X			X	X
Invoke	Procedural	Report Error	X		X		X	X	X
Invoke	Provide Instructions	Additional	X					X	
Invoke	OPDU		X			X			
Return Result			X		X	X	X	X	X
Return Error				X	X	X	X	X	X
Reject			X	X	X	X	X	X	X
X = Message Used									

Return Error
Reports that an operation was not successfully completed.
Reject
Reports the receipt of a message causing a protocol error at the transaction or component level.
The SCP and the SMS must exchange information and control each other's operations. Two classes of messages are defined in the SCP/SMS interface: measurement data and status data[9].

---

[9] SCP-SMS Generic Interface Specification, Bellcore, TA-TSY-000365.

- Measurement Data

  The SMS collects data from the SCPs about performance and traffic. The SCP software must be designed to provide the SMS with separate sets of measurements for the SCP operating system, the SCP support software, the signaling traffic between the Signaling Transfer Points (STPs) and the SCPs, and each application implemented at the node.

  Measurement messages record hardware and software failures, and traffic in and out of the SCP. The SCP node transmits the measurements collected by the applications. The SMS periodically requests each SCP node to transmit one interval of data collected since the last request. The SMS validates and stores the data.

- Status Data

  The SMS acts to minimize SCP performance degradation caused by traffic overload (and thereby minimize data loss when either the SMS or the SCP fails) and to maintain the consistency of the SMS and SCP database. The status messages from the SCP to SMS report on application availability, SCP overload status and the assumption and later release of its mate's traffic by an application.

Table 3 shows the generic messages sent between the SCP and the SMS. For example, the Good Night and Good Morning messages are used for shut-down, start-up, and link recovery procedures. The messages are used by the SMS to prompt for information, and by the SCP to provide information.

Table 3 (Page 1 of 2). Generic Messages Between SCP and SMS[10]		
Messages	**SCP to SMS**	**SMS to SCP**
Retrieve Application Measurements		X
Report Application Measurements	X	
Good Morning 1	X	
Good Morning 2		X
Good Morning 3	X	
Good Night		X
Retrieve Indicated Status of Application		X
Response to Retrieve Indicated Status of Application	X	
Report Indicated Status of Application	X	
Site to Site Confirmation	X	X
Tell Node Update Complete		X
Report Measurements Space Low	X	

---

[10] SCP-SMS Generic Interface Specification, Bellcore TA-TSY-000365.

Table 3 (Page 2 of 2). Generic Messages Between SCP and SMS[10]		
Retrieve Node Measurements		X
Report Node Measurements	X	
Report SSP List Error	X	
Report Scheduled Network Management Results	X	
X = Message Used		

Triggers

Trigger tables determine when an SSP should send a query (for example, PROVIDE INSTRUCTION) for a certain type of call to an SCP. The processing capability of a possible future generic trigger table as currently described by Bellcore is considered in *Service Switching Point 2— SSP/2 Description*, Bellcore SR-TSY-000782. Craft persons update trigger tables via the normal operations and maintenance interfaces.

2.8 Databases

2.8.1 Size

The size of the SMS and SCP application databases was determined for the five described services and selected European, American, and Pacific countries. This data is implementation-specific, therefore, it has not been published in this report.

The application database size in the SMS and SCP depends on the number of service subscribers or users, and the record length. With PVN and AWC, service subscriber network configuration data also effects the database size. The number of service subscribers or service users was derived from the total number of telephone subscribers expected in the investigated countries for the projected time periods. The record length basically depends on the data elements required to provide a service (for example, routing and number translation data, indicators for validation, and feature data). It is assumed that the record contents are very similar in the SCP and SMS, merely augmented by subscriber administrative data for the SMS.

The record sizes within a database can vary considerably according to the functions and features a subscriber uses for a particular service. Therefore, not only average but also maximum database sizes were calculated. The SMS and SCP must be able to support a variety of database sizes, ranging from megabytes to several gigabytes.

2.8.2 Database Integrity Requirements

Self-checking procedures must be provided to detect errors in the database records. The SCP must perform periodic consistency checks with the SMS to ensure that SCP data is correct. The SMS must have a backup copy of all data provided to the SCP, to be used for consistency tests and to restore the SCP databases where data is destroyed or the SCP fails.

Data integrity and consistency must be guaranteed for a series of database updates. Either all updates or no updates should occur. The transaction system has to provide data backout and recovery mechanisms, to ensure correct updates in case of system or application failures. For example, a GNS subscriber wants to change his routing from all day via long distance carrier A, to 8 a.m. until midnight, and add long distance carrier B from midnight to 8 a.m.. He wants this change to take place in a week. There are at least three updates required:

1. Update the current version record specifying carrier A from all day, to 8 a.m. until midnight.
2. Add a new version record specifying carrier B from midnight to 8 a.m..
3. Add the changes to the store and forward file.

The service subscriber expects that if he is not notified of errors, his changes will become active at the correct date, especially because his changes were entered some time earlier.

This is basically the concept of transaction integrity, which implies that if the operating system or application fails, the transaction system will continue to accept responsibility to complete the work as soon as the failing condition has been fixed. In addition, transaction integrity implies that if there are any messages to the subscriber, they will be delivered once and only once.

All update requests sent from the SMS to the SCP have to be queued in the SMS. In case of a link, application, or node failure they can be transmitted again once the failing component has been restored.

2.8.3 Location of Databases

IN databases can be replicated in more than one SCP, or distributed among several SCPs (each SCP has only part of the database). (See Figure 7.) No part of the IN database is located in the SSP.

The SCP part of the database is service specific. It resides on disk (backup copy) and may also reside in memory for performance reasons.

The SMS part of the database is service specific. It resides on disk and is referred to as the master database.

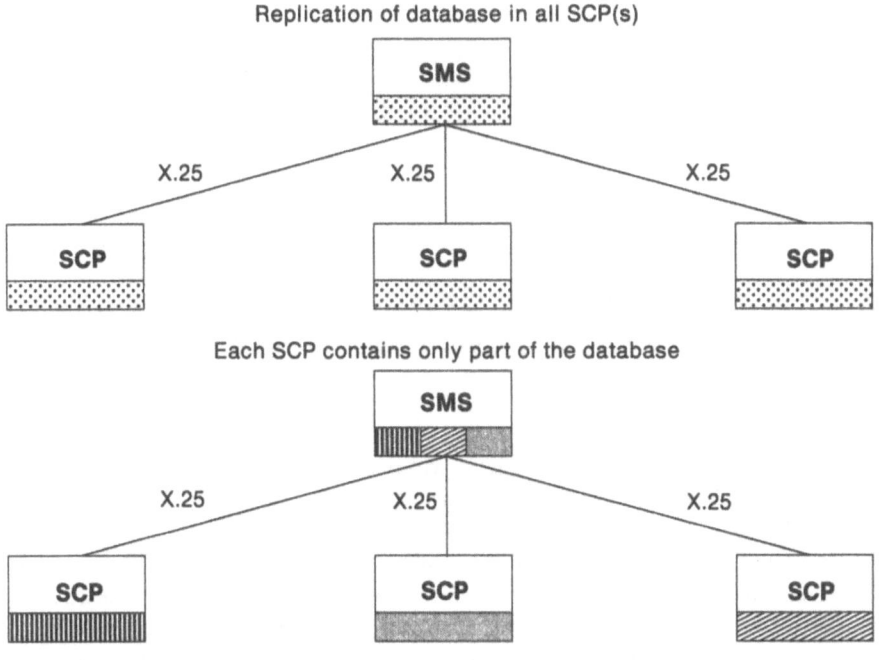

Figure 7. Database Distribution

2.8.4 Database Administration

Initial Loading

The SMS must be able to perform an initial load of its own and all relevant SCP databases. Initial loading requires a high degree of parallelism because the database can be very large. A main process in the SMS drives N subprocesses (where N is the number of physical devices); it reads from the initial load tape, uses a hashing function to determine the destination device and location within the device, and then passes the record to the subprocess associated with the destination disk. The main process operates faster than the subprocesses, so that all disks are updated at full speed.

The SCP databases can be loaded in two ways; both start with the SMS producing load file magnetic tapes:[11]

- **Loading via an X.25 links**
  1. The load file is sent to the SCP using file transfer capabilities.
  2. The application software is started via a command from the SCP local terminal.
  3. The application's initial state is "disabled", which means normal SMS/SCP application connection (via the SCP node) is not established.

---

[11] SCP-SMS Generic Interface Specification, Bellcore TA-TSY-000365.

4. The SMS load file is loaded.
5. The application generates a response file, containing one response message for each message in the SMS load file. The response file is written onto magnetic tape and transferred to the SMS (via file transfer or manually).
6. The SMS processes the response file and marks successfully loaded records as "active". Unsuccessful records are marked "failed".
7. The SMS prints a report listing the records that could not be loaded.
8. The operations personnel examine the report and take appropriate actions.

- **Local loading via tape**
  Because some databases are very large and take a long time to load, loading is made easier by transporting the database on tapes to the SCP and loading it locally, via magnetic tape stations.

Record Update

There are three types of database updates: addition of a new record, deletion of a stored record, and change of fields in a record.

Having received all update data, the SMS queues it for editing and validation (done by special checking and validation programs). Once this is done, the service master database in the SMS is updated, and the updates are placed on a store and forward file to update all related databases. This process is done asynchronously to the subscriber input action under control of the SMS, using transaction processing between the SMS and the SCP(s). The database updates can be initiated by:

- The service subscriber filling out a service request
- The subscriber calling a network operator service center to update his service parameters. The network operator personnel update the service parameters using a data terminal connected to the SMS, via a dial-up connection or a dedicated line.
- The service subscriber may update his own service parameters directly from a terminal, via a dial-up connection to the SMS (to which he has security protected access). The subscriber can specify when the updates should become effective.

2.9 Transaction Rate Estimates

2.9.1 Methodology

Bell Atlantic, IBM, and Siemens developed a methodology to determine the transaction rates and storage capacity requirements for the IN elements.

This methodology is based on demographic data, interviews, and detailed knowledge of telecommunications markets worldwide. Due to the sensitivity of this implementation-specific data, it has not been published in this report.

All IN elements must be engineered to handle peak transaction rates (traffic). The most critical element in the IN architecture is the SCP because it acts as a focal point for transactions generated in the network and concentrated through the CCS7 network. (**Note**: The STP handles the transactions for several SCPs, but the processing done for each transaction is less.) In addition, the SCP receives transactions from the SMS as a result of service subscriber data updates. The SCP generates transactions to the SMS in the form of measurements, traffic, and billing data. The following methodology was developed to determine transaction rates between the IN elements.

2.9.2 Transactions Between SMS and its Users

The network operator personnel and (if authorized) service subscribers are the SMS users. A transaction between an SMS and its user is the action(s) that takes place from when the user presses "Enter", until a response is received at the user's terminal. (For example, a change of menu, calling a help option, scrolling through call statistics, or changing service parameters.)

The study developed a method of estimating the number of service subscribers per service and per country. The annual number of logon-sessions and transactions per session was assumed, based on network operator data and experience. With this information, the study group estimated the peak transaction rate between the SMS and its user.

2.9.3 SMS-SCP Transaction Rates

The number of SMS-updates (which change service subscriber data) largely determines the number of transactions from the SMS to the SCP. Transactions from the SCP to the SMS contain information like measurements, traffic data, and billing data. To determine peak transaction rates, the study group estimated the amount of data, average message length, and the frequency with which this data is sent from the SCP to the SMS. (The number of subscriber data updates has already been determined for calculating SMS user updates.)

With this information, the study group estimated the SMS-SCP peak transaction rate. This rate was smaller than the SMS-user transaction rate and negligible compared with SSP-SCP transactions.

2.9.4 SCP- SSP Transaction Rates

To determine the transaction rates between the SSPs and an SCP within a country or region, it was first necessary to determine the annual number of calls associated with a specific service. The number of service calls was distributed to estimate the peak transaction rate. From this information, the study group determined the maximum number of transactions the SCP must be able to handle. Based on publicly-available telephone statistics (such as UIT) and other sources, the number of telephone subscribers, the growth in this number, and the number of calls per year was estimated.

To calculate the number of calls associated with a specific service in a given year, the study group used these figures and the companies' knowledge of the volumes of actual and anticipated service calls. This number was calculated as a percentage of the total number of telephone calls per year.

A formula was developed to calculate the peak transaction rate for a service. This peak rate depends on factors like the number of transactions per service call, traffic during busy periods, and the number of busy days per year. With this information, the study group derived service and country specific peak transaction rates for the period studied.

2.10 Measurement Requirements

The LATA Switching System Generic Requirements (LSSGR) categorize the traffic measurements into two types, based on the method of data collection. The counts required are **usage counts** and **peg counts**.

Usage is the duration of a specified condition, normally the busy condition of a server (a system component that performs operations required for the processing of a call). It is collected and accumulated over time intervals specified for the type of server.

Peg counts measure the number of occurrences of events. Specifications of individual peg counts are in terms of system recognition and response. Peg counts should be 100% accurate.

2.10.1 SSP

Note: This section is based on information from "Service Switching Points", Bellcore TR-TSY-000064, FSD 31-01-0000.

A number of peg and usage counts are provided within the specified recording periods:

- 30-minute Administration Reports, which should be produced on the hour or half hour.
- Hourly Maintenance Report, which is a subset of the daily maintenance report and which should be accumulated every hour on the hour.
- Daily Maintenance Report, which should be collected every 24 hours, starting and ending at midnight. It should be possible to schedule the time of output for the Daily Report for any time up to 24 hours after the end of the accumulation interval.
- Five minute Network Management Report, which should be produced with the specified measurements so that the network performance can be monitored.

Some **maintenance** measurements are:

- Failure to obtain Automatic Number Identification (ANI) on a Centralized Automatic Message Accounting (CAMA) Trunk (Access Tandem/SSP only)

- Failure to receive second signaling stage on an exchange access trunk (AT/SSP only).
- Call processing failure before initial query
  An SSP call is terminated due to hardware or software initialization, or due to a failure in the normal call processing routine before sending the query message to the SCP. Any call processing failure that can be detected by the SSP should be counted under this category.
- Call processing failure after initial query
  An SSP call is terminated due to a hardware or software initialization, or due to a failure in the normal call processing routine after sending the query message to the SCP. Any call processing failure that can be detected by the SSP should be counted under this category.
- Resource unavailable before initial query
  A call fails because a resource, which is normally provided by the SSP, is unavailable due to maintenance or engineering reasons. Keep a separate count for each resource.
- Resource unavailable after initial query
  A call fails because a resource, which is normally provided by the SSP, is unavailable because of maintenance or engineering reasons. A separate count should be kept for each type of resource.
- NM control blocks call
  The call is blocked by network management controls on the calling number.
- Signaling failure time-out at the SSP
  The SSP times out while waiting for a reply from the SCP.
- Invalid command message
  The SSP receives a response from the SCP that is indecipherable, or has invalid data.
- Invalid command sequence
  The SSP receives a response from the SCP that contains an incomplete or out of sequence set of commands.
- Return Error or Reject Message
  The SSP receives a Return Error Message or a Reject Message in response to a query.
- Abandon before outpulsing
  An on-hook is received from the calling party before the SSP seizes an outgoing trunk.
- Abandon after outpulsing
  An on-hook is received from the calling party after the SSP seizes an outgoing trunk, but before the call is answered.
- All Interexchange Carrier (IC) trunks busy
  A call cannot be completed because there are no trunks available on the final route to the IC.
- All network operator trunks busy
  A call cannot be completed because there are no available trunks on a call to be completed via the network operator.

Base count measurements should be made so the volume of service calls can be monitored. The SSP should count all service calls originating in the SSP switch that have reached the dialing complete stage, and all service calls that have been received from a non-SSP switch.

2.10.2 SCP

Note: This section is based on information from "SCP-SMS Generic Interface Specification", Bellcore TA-TSY-000365.

The SCP node transmits the set of measurements collected by the node and stored by the SMS. The measurements contain data on the SCP hardware, applications, system and node software, traffic, and the CCS7 network performance.

The SCP collects and stores node measurements in 30-minute intervals. The SMS periodically polls the SCP node for these measurements.

The SMS sends a Retrieve Node Measurements message to the SCP node where it triggers the generation of a "Report Node Measurements" message which is subsequently sent to the SMS. This message contains measurements that have been stored at the SCP since the last retrieve measurements message from the SMS. The Report Node Measurements message should be no larger than 4000 bytes. If the data measured for the node during the interval is greater than 4000 bytes, it is indicated by a flag in the SCP response message. Thus, additional data is retrieved by the SMS by sending another Retrieve Node Measurements message. The SCP node should be large enough to maintain three days of measurements. If, at the end of three days, the SMS has not collected the data, the SCP node should destroy the data of the oldest 24 hours and begin to store new measurements.

Application measurements are done in 30-minute intervals that begin on the hour. (These are the same 30-minute intervals as those in which the node measurements are collected.) At the end of the interval the application passes the measurements to the SCP node where they are later retrieved in the same way as node measurements.

The SCP collects network management measurements in five-minute intervals and sends them to the SMS as a scheduled report. These reports are sent to the SMS by the SCP at the end of every five-minute interval, provided there is measurement data to be reported. Once the SCP has received an acknowledgement from the SMS for successful transmission of the network management measurement report, the data can be destroyed at the SCP.

2.10.3 SMS

Note: This section is based on information from "Service Control Point Node—Generic Requirements", Bellcore TA-TSY-000029.

The SMS collects all measurements from a 24-hour period, starting at 12:01 a.m. local time. The SMS requests these measurements from the SCP as required. (They are stored in a measurement database.)

The measurements are for each 24-hour period, are retrievable at the end of the period, and remain so for 72 hours. Once the measurements are received by the SMS, they are no longer available for retrieval at the SCP.

The SMS can measure the traffic load in the service modification process of a service subscriber. This information is stored in the traffic statistics master database in conjunction with the traffic data supplied by SCP and SSP. The service subscriber uses the database for online traffic statistics queries, or to generate reports on the volume of traffic for his service.

2.11 Performance and Dynamic Requirements

2.11.1 Response Time

Service User
The response time or call setup time (CST) is the period between completion of dialing until receipt of audible ringing.

A network CST is reduced by introducing CCS7, but may be increased by the introduction of the IN architecture. This is explained in 4.2.1, "Performance and Throughput Requirements".

Service Subscriber
This response time is the period between pressing ENTER and receiving a response. The response time[12]:

• For a minor change of menu (scrolling, help, and so on), should average one second, and in 95% of cases be under three seconds.
• For an update of service information should be within 15 seconds (but varying according to the complexity of the update).
   These figures do not include transmission time.

---

[12] Plan for the Second Generation of the Intelligent Network, Bellcore SR-NPL-000444.

2.11.2 Response Times for IN Elements

SSP Response Time: The SSP response to a service user request should follow normal requirements. The SSP waits three to four seconds after a query is sent to the SCP before timing out.

STP Delay Time: The message delay is measured from the time the STP receives the last bit of a TCAP message on an incoming signaling link until the last bit of the message is transmitted on an outgoing signaling link. The delay of a message through an STP depends on factors like message length, and whether it requires global title translation.

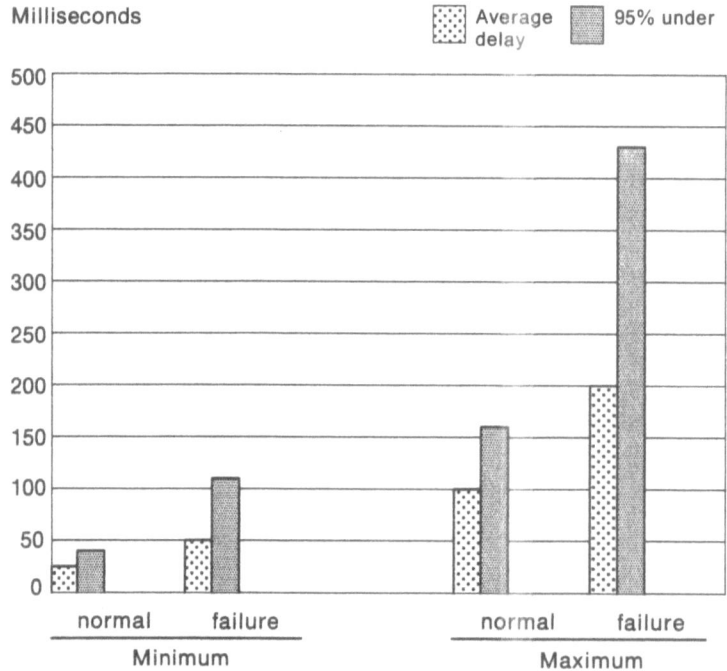

Figure 8. STP Message Delay[13]

Figure 8 shows the maximum permissible delay times in an STP when all messages are 15 octets long and no global title translation is required (marked with "Minimum" in the figure).

---

[13] Data from Signaling Transfer Point Generic Requirements, Bellcore TR-TSY-000082, Issue 2, June 1987.

Figure 8 shows the maximum permissible delay times in an STP regardless of message length or whether global title translation is required (marked with "Maximum" in the figure).

There may be more than one STP transfer between an SSP and an SCP, depending on the structure of the network.

SCP—SMS Messages: The SCP node sends a message to SMS in two situations:

- On the receipt of an SMS command which requires a response
- On the occurrence of an event at the SCP of which the SMS should be notified via a status message or report.

Average response times for generic messages received from an SMS should be no greater than eight seconds. The response for 99 percent of SMS messages should be within 16 seconds[14]. The service specific response time usually covers the following:

1. The SCP/SMS interface in the SCP receives an update message.
2. The message is sent to the update application in the control host, which is the primary application host processor.
3. The database in the control host is updated.

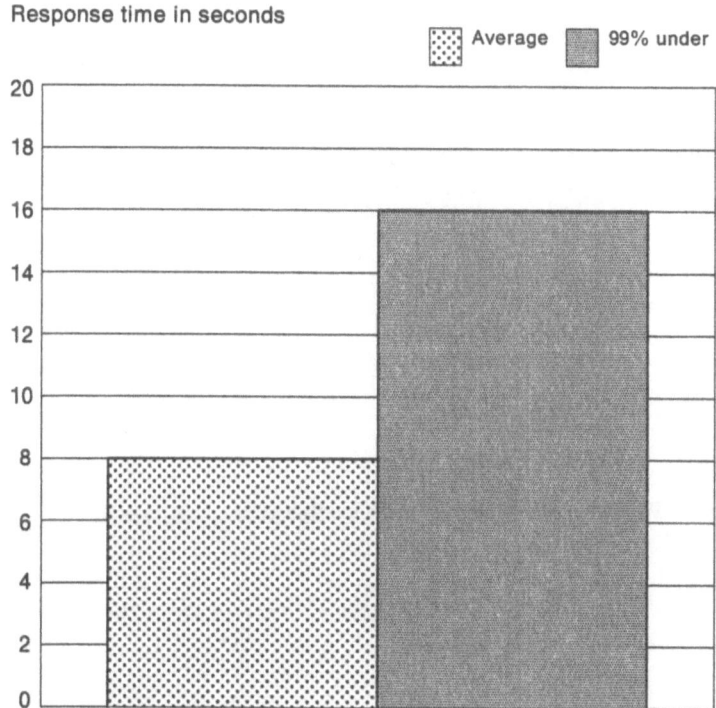

Figure 9. SCP Response Time of SMS Generic Messages

---

[14] SCP-SMS Generic Interface Specification, Bellcore TA-TSY-000365.

4. An update message is sent from the control host to all other hosts.
5. The update is performed in all subordinates, and each other host sends an "update successful" message to the control host.
6. When all updates are completed, the control host sends an update final notice to the SCP/SMS interface.
7. The SCP/SMS interface notifies the SMS that the update is successful.
Figure 9 shows the response time elements; which is the same for all selected services.

SCP–SSP Messages: The SCP responds to an SSP message within a prescribed time. Figure 10 shows the response times for a "Provide Information Request" for the individual services.

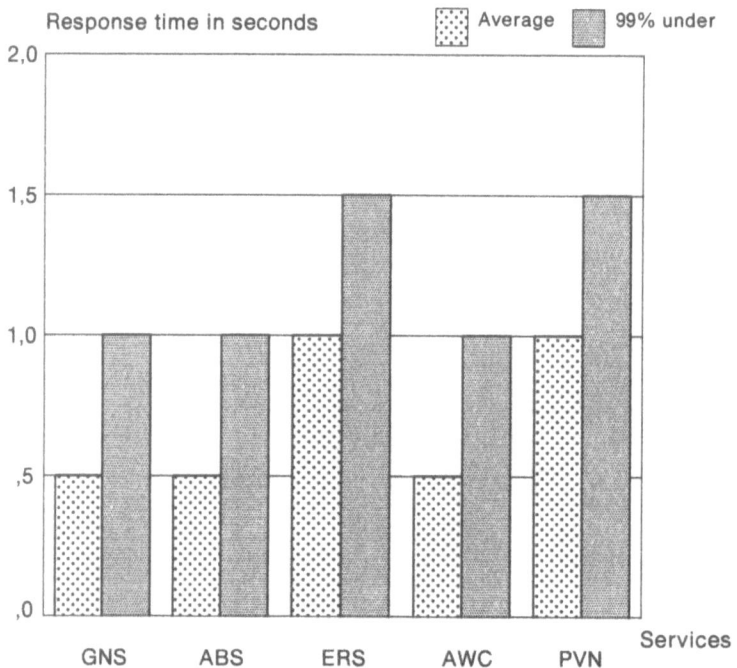

Figure 10. SCP Response Time for SSP Messages (Service Specific)

2.11.3 Overload Handling

The initial request for overload handling is described below.

Message Transfer Part (MTP) Level
The CCS7 overload handling must be applied to IN/1.
If the CCS7-MTP implementation is overloaded, the source of messages must reduce message creation (CCITT international version), or messages are discarded according to their priority (US version).

If the CCS7 user (in this case, the SCP application) is overloaded, overload messages may be sent via CCS7 to reduce traffic production.

Application Level
The SCP node detects an overload by measuring the delays between receiving a query and returning a response, or by measuring the length of the incoming messages queue. When the SCP node determines it is overloaded, it notifies the application. The application responds to help alleviate different levels of overload.

The applications take the following call processing actions:[15]
Level 0   No overload.
Level 1   The SCP informs the SMS, other administrative centers, and all applications of its overload level. The node assumes that the administrative system will suspend all low priority transactions.
Level 2   As for level one, but with normal administrative messages.
Level 3   Implement automatic call gapping (ACG) at level x, where x = 1 through 8.
Level 4   Stop collecting application measurements. Reject query messages except conversations with permissions and provide additional instructions.
Level 5   Reject SSP error messages without termination indicators.
Level 6   Reject SSP error messages with termination indicators.

Priority
Messages received in an SCP may come from different SSPs (and SMS) and Operator Services Systems (OSSs) and from different applications within these components.
    A message must have the capacity to be prioritized according to country-specific needs. For example, OSS messages generally have higher priority than SSP messages and ERS messages have the highest priority.
Note: CCS7 messages cannot be prioritized. However, the actions taken in response to a CCS7 message must be able to be prioritized according to country-specific requirements. For example, the action taken on an ERS message should be given greater priority that the action taken on a GNS message.

---

[15] Service Control Point Node - Generic Requirements, Bellcore TA-TSY-000029.

2.12 Availability Requirements and Objectives

The requirements and objectives of availability are based on Bellcore requirements[16].

SSP: The SSP function operates on a network switching element, therefore the maximum downtime must not exceed three minutes per year. (The introduction of a new generic software version should not make the network switching element unavailable.)

STP: Refer to "Signaling Transfer Point Generic Requirements", Bellcore TR-TSY-000082.

SCP: "Service Control Point Node—Generic Requirements", Bellcore TA-TSY-000029 states:

It is desirable that the SCP is available 24 hours per day, 7 days a week. Actual availability requirements allow for minimal downtime as discussed below.

The availability objectives for the RBOC CCS7 network state that the downtime from signaling endpoints to endpoints should be no greater than approximately 10 minutes per year. This downtime objective includes the transmission path across the CCS network as well as the CCS7 protocol interface at the SCP. It excludes the actual downtime at the SCP caused by other hardware or software failures.

The availability requirements for the SCP itself will be determined by the applications that are in service at a particular site. The SCP (node plus application) should satisfy the most stringent requirements of the applications. The availability requirements specified by an application define the total time it should be possible to access the application data from the network. Because SCP node availability is essential for data access, the availability specification incorporates node availability as well as application availability. It is expected that maximum allowed downtimes will be in the order of 10 to 20 hours per year for a single site. Suppliers need not design to meet downtime requirements of less than 10 hours per year.

SMS: The SMS should be available 22 hours a day, every day of the year. Two hours of each day are used for SMS maintenance.

---

[16] General reliability and quality requirements for switching systems, including the SCP, are found in Reliability and Quality Switching Systems Generic Requirements, Bellcore TA-TSY-000284 and LSSGR LATA Switching Systems Generic Requirements, Bellcore TR-TSY-000064.

2.13 National Considerations

The International application of IN services is important for smaller countries, especially in Europe, for example, so that a GNS user in Austria can dial a GNS number in France.

The CCS7 signaling network is generally a national network, therefore service interworking between countries is usually performed from ISDN User Part/Telephone User Part (ISUP/TUP) to ISUP/TUP.

The possibility of dialing into an SMS and accessing subscriber data is a potential security risk. Every effort must be made to minimize the potential for unauthorized use of the network, for example, by establishing a special logon procedure, whereby:
1. The service subscriber dials the SMS number and is connected to the SMS, and supplies a password.
2. The subscriber hangs up.
3. The SMS calls the subscriber back.
4. The subscriber gives a new password.
5. The subscriber logs on using his user ID, account number, and service identification.
6. The session between the service subscriber and the service logic is established.
If a service function is centralized at only a few SCPs, the service is susceptible to terrorist or enemy attacks. The IN architecture is so flexible that it can be configured in order to minimize this risk.

2.14 Future Considerations

Billing information is normally collected in the SSPs. In the future it may be desirable for the SCP to collect billing information. This may lead to a substantial increase in the TCAP messages flowing from the SSP to the SCP.

As technology improves, the IN services described may improve their interactive capabilities (for example, through voice recognition) so that a POTS user without MF signaling can use the interactive features.

Services related to voice calls should be extended to ISDN calls, as ISDN becomes increasingly available.

The trigger table data should not be defined in the network switching element generic software. It should be administered by a network operator through an operations system so that no changes to the generic software are necessary when updates are made to the trigger table. Table 4 shows an example of a possible trigger table.

Table 4. Trigger Table Example[17]

Trigger		Point in Call	Message Address—Global Title		Service Key
Type	**Value**		**Type**	**Value**	
Call Address	800		1	Call Address	Call Address
	900-123		1	Call Address	Call Address
	900-742		1	Call Address	Call Address
	411		7	Call Address + LATA	Call Address
	911		8	Call Address + ANI	ANI
Calling Address	201-758	Dial Tone	2	ANI	ANI
	201-740	Dial Tone	9	Calling Address	Calling Address
Originating LATA	224		3	LATA + ANI	LATA
Carrier	127		5	Carrier	Carrier
	288	Call Address	6	Carrier + Call Address	Carrier
Service Code	74		8	74 + ANI	Service Code
Facility Code	53		10	53	Facility Code
Prefix	0	Call Address	2	ANI	ANI

Translation Types (Sample Values)
1. Incoming Call Management
2. OLNS
3. LATA—Defined Services with ANI
4. LATA—Defined Services
5. Carrier—Defined Services
6. Carrier—Defined Services with Call Address
7. Service Code with LATA
8. Service Code with ANI
9. PVN
10. Facility Code

---

[17] Service Switching Point 2—SSP/2 Description, Bellcore SR-TSY-000782.

Chapter 3. Network and Service Management

3.1 Overview

One of the goals of the Intelligent Network is to open a programming interface for the development of services by the PTT or third parties, rather than the network element provider. Customers will have direct access to the IN to establish, change, and modify their services. Normal call processing by the exchange is suspended until a centralized intelligence is consulted for call-processing instructions.

This requires stability, security, and sound performance in the network management system. Parts of any one service may exist in each of the main IN components: the SSP, SCP, and the SMS.

A successful network and service management system (NSMS) will include:

- Integration (a comprehensive, end-to-end view of the network)
- Open Systems Architecture (published and accepted interfaces)
- A universal Man-Machine interface.

The development of an appropriate network and service management system should be within the time-frame that many PTTs are considering for the introduction of the IN itself, and an effective system is important for the IN's long-term success. Centralized operating system functions are essential to manage the end-to-end delivery of services, from the service provider interface to the SMS system to the service user's access terminal.

The placement of service functions depends on the operating environment of each IN, but the most sophisticated IN will demand a separate operating system, to provide a central point for customer contact and problem determination.

This chapter identifies and assesses the requirements of a Network and Service Management System (NSMS) for an Intelligent Network, and identifies the problems of and opportunities for development in both the short and long term.

3.2 Methodology

This chapter establishes the attributes an NSMS must have in order to support the
IN and to assure high quality global provisioning, execution, and delivery of IN
services, at acceptable costs. Basic operation, administration and maintenance
(OA&M) functions required for the network are also identified. These are
network-wide and include:
* Configuration management
* Monitoring and control
* Maintenance
* Administration
* Testing.
Given these requirements, the existing (or planned) operations, administration and
maintenance systems designed to support the various networks associated with IN
are considered. These include the Bellcore Signaling, Engineering and
Administration System (SEAS) for the CCS7 network, and Bellcore Service
Management System (SMS) for management of the Service Control Points (SCPs).

Finally, the requirements of the operations, administration and maintenance
systems, as individual and IN-wide functions, are compared, and trends for a future
NSMS are identified.

3.2.1 Summary of Findings

* IN implementation will proceed relying on existing or planned sub-network
  level systems regardless of whether an adequate Network and Service
  Management System (NSMS) exists.
* In the long term, an **integrated** set of **centralized** functions is crucial to the
  successful implementation of IN services.
* For a successful NSMS, it is vital to define a migration path from the existing
  network administrative systems to:
  − A comprehensive and integrated set of IN functions
  − An Open Systems Architecture with published and accepted interfaces
  − Early IN implementations and IN/2.
* A considerable effort is required to provide a comprehensive definition of the
  NSMS scope and the required international standards. Standardization
  activities are currently taking place inside CCITT under the title of
  Telecommunication Management Network (TMN).

3.3 NSMS Requirements

This section describes the basic requirements used later for assessing the potential in NSMS development.

3.3.1 NSMS Global Qualities

It is often said that the success of the IN architecture will largely depend on the success of network-wide service management. This suggests that the efficiency of administering and maintaining services will be equally important to, if not more importantant than, the efficiency of the service itself. This is easy to understand, because if a service requires network-wide "orchestration", the service must be established, and maintained before a user can employ the service.

Based on this premise, the essential quality of an NSMS is the ability to direct the operational, administrative, and maintenance activities associated with a service, comprehensively and efficiently, at all network elements (for example, the SSPs and SCPs associated with the service). This must be achieved using standard protocols.

A telecommunications network is a collection of nodes and links that communicate by defined sets of formats and protocols. Within the network there are usually three layers:

- The transmission layer consisting of transmission systems (for example cables, radio links and their related technical equipment).
- The switching layer consisting of switching nodes with generic and application software and data.
- The service layer (distributed among the switching network elements) consisting of special hardware such as SCPs, announcement machines, and their application software and data. The domain of network management is usually the domain of one Telco or PTT.

An NSMS should encompass these three network layers, as a service could affect them all. The introduction or monitoring of a service could require provisioning to, or access to, all network elements (NEs).

Figure 11 depicts the current status of Operations Support Systems deployed in the United States for the IN.

The links to support systems are either dedicated links, with vendor specific formats and protocols, or X.25/BX.25 links.

3.3.2 Basic Operations, Administration and Management Functions

This section highlights the basic functions of a telecommunications network management system.

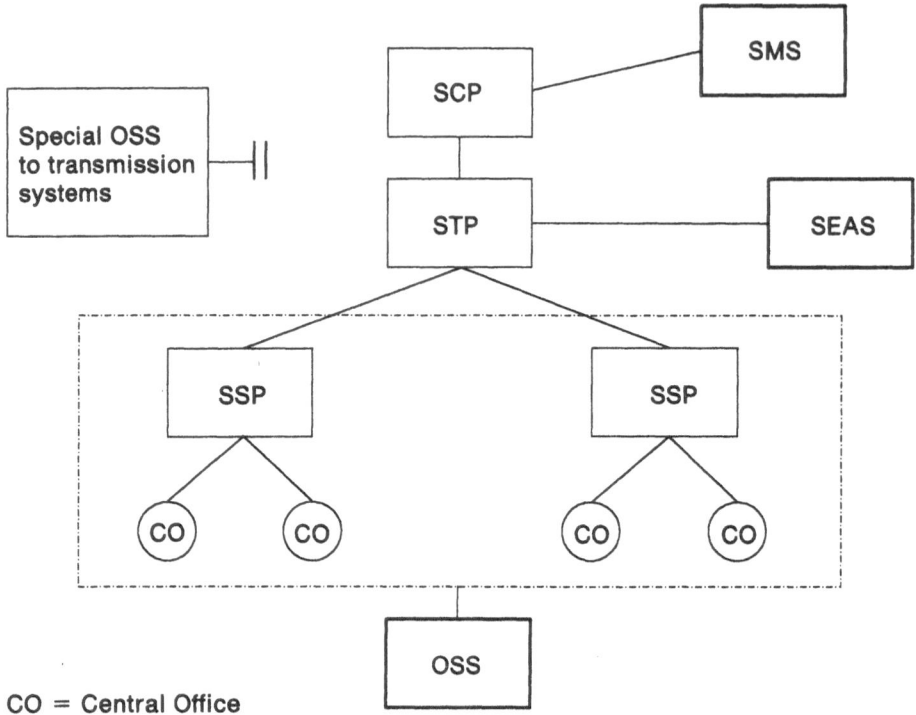

Figure 11. Operation Systems within the United States' Intelligent Network Operations

Configuration Management
Configuration management is the provision of functions and databases to maintain the physical implementation of the network. The databases contain information about the network elements and status attributes like "operational" or "faulty".

Configuration Management is used for network planning and updates, and is necessary for diagnostic and maintenance control. Information stored in network configuration data bases should be as accurate as possible—and thus should be maintained by interactive or automated configuration management tools.

Monitoring and Control
This function includes all measures taken to keep the network operational. Examples are:
- A system interface through which network elements report their actual status to the operations system in real time (for example alarms, traffic counters, billing information, and overload indicator) and the operations systems react.
- Functions built into the network elements for real time recovery provisioning and related reports.
- Automated facilities for online reaction to network element-alarms. Semi-automated functions to allow craftsmen to control the network via the human interface.

- Performance management.
- Diagnostics, online traces and fault management.

Maintenance
The measures to repair non-operational NEs, or to perform preventive actions. For example, disconnecting NEs, repairing them, and bringing them back to operation, or distributing software updates to the NEs.

Administration
This function includes:
- Security management (access, passwords, and encryption)
- Directory management (for example, names and addresses for customer lines)
- Management of semi-permanent data (for example, routing and translation tables and network and NE system parameters)
- Service data management.

Testing
A set of test facilities are provided for both physical facilities such as transmission equipment and logical entities such as database updates.

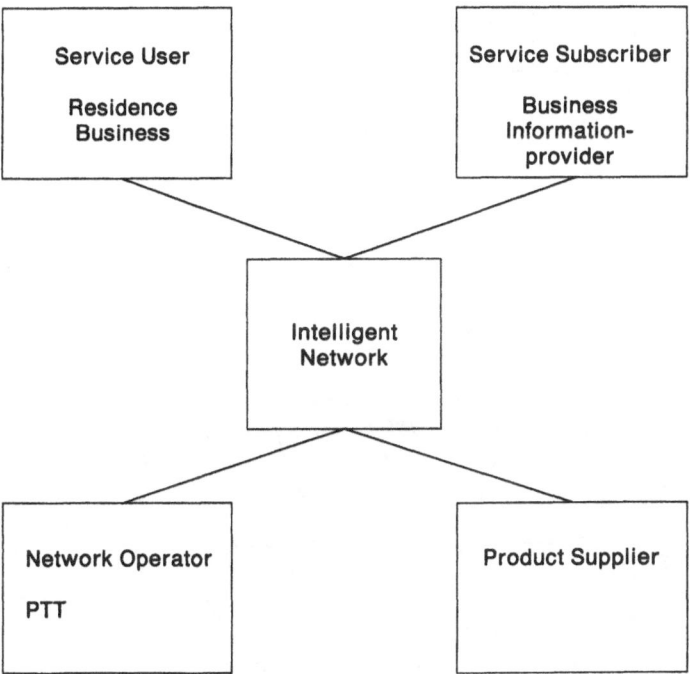

Figure 12. Network User Categories

3.3.3 Network Management Users

It is important to realize that the users of the NSMS will be all the active
participants in the IN, as depicted in Figure 12. This fact becomes more relevant
when future NSMS opportunities are discussed in this chapter. Thus, there will be
more than one user involved in the task of network management, including:
- The network operator (as expected)
- The service subscriber (for example, to update his service data)
- The network product supplier (for example, for maintenance)
- The user himself (for example, for private virtual network services).

Most PTTs currently manage their existing network layers via separate and
independent systems. The USA, for example, uses a host of systems known as
Operations Support Systems (OSSs) for the switching layer, the Signaling
Engineering Administration System (SEAS) for the CCS7 transmission layer, and
the Service Management System (SMS) for the service layer (see Figure 13).
Figure 13 also indicates the users affected by each management system.

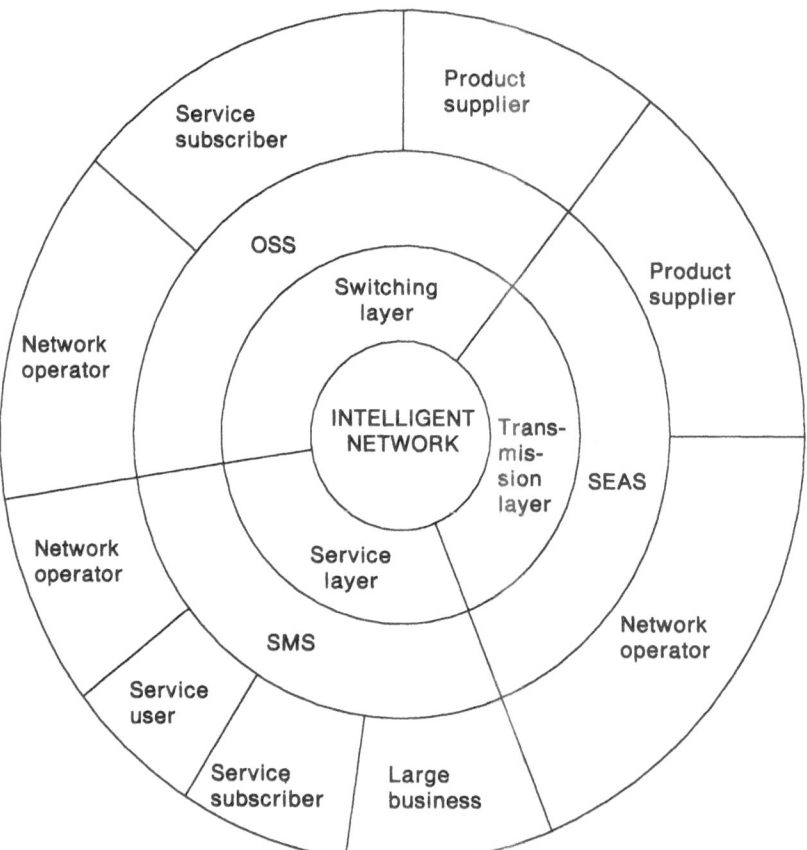

Figure 13. Operations System Users

3.4 Switching Layer and the OSS—an Example

The Operation and Maintenance Data Communications Systems (OMDS) is an OSS developed by Siemens to control the digital switching system (EWSD). It is adaptable to the national requirements of the PTTs.

Configuration Management
Tools for producing software-updates for EWSD and OMDS itself are available on the host system BS 3000 (MVS, 7.800). To update the software, one of the following changes is performed:
* Patches
* Exchange of a software-module
* Production of a new software-release.

Monitoring and Control
To keep the system operational, OMDS offers these measurement categories:
* Base measurements (equipment quantities such as lines and trunks are listed)
* Service measurements (reporting on call setup troubles, rejected call attempts, canceled calls)
* Performance measurements (trouble counts, reports on the recovery actions of components such as the Line Trunk Group and the Digital Line Unit)
* Country specific measurements.
Traces for system components (on different levels of detail) can also be obtained.

Maintenance
Alarms report on defective EWSD components. The components can be automatically disconnected from the system, separately tested with diagnostic software, repaired, and returned to operation.

For reliability reasons, alarms are transferred via additional (separate) transmission paths and are indicated to the Network Operator Personnel on a special panel.

Administration
Access to administrative functions may be controlled by the requirement of user identification and password. Commands can be grouped together and a password assigned to such a group.

The user interface is designed according to the Man-Machine Language (MML) or the Extended MML, respectively, recommended by CCITT.

In the administration of subscriber data, information like call number, type of device (rotary dialing, push button dialing), and whether a private meter is installed can be entered, changed and deleted.

Testing

Before new components or software updates become operational, a thorough test must be performed. Hardware components can be tested with diagnostic software (see "Maintenance"). To introduce software releases, utility programs are available to integrate existing data (for example, subscriber descriptions) in the new release. Depending on the software modification, the data may have to be converted before being integrated.

3.5 Signaling Network and the SEAS

3.5.1 Management Functions Related to the Signaling Network

The signaling network consists of:
* Service Switching Points (SSP)
* Signaling Transfer Points (STP)
* Service Control Points (SCP).

Signaling System No. 7 comprises:
* Message transfer part (MTP) containing:
  Level 1: Signaling data link (Q.702)
  Level 2: Signaling link (Q.703)
  Level 3: Signaling network functions and messages (Q.704)
* Different user parts (Level 4).
  For IN/1, the relevant functions are the Signal Connection Control Part (SCCP), the Transaction Capability Application Part (TCAP), and after its completion the Operations and Maintenance Applications Part (OMAP).

There are signaling network management functions in the MTP in the CCS7, or provided by SEAS. These functions may be network element- and implementation-specific.

The early releases of the Signaling, Engineering, and Administration System (SEAS) provide data collection and a simple data reporting function for network management personnel[18].

Later releases will provide more sophisticated techniques, evaluation and analysis of data. The SEAS may be situated locally within a Signaling Engineering and Administration Center (SEAC).[19]

SEAS is directly connected to the STP. Using a Bellcore network structure, where the SSP and SCP are connected to STP in a strictly defined way, the reports from SEAS give information about the node part of the connected SSP and SCP (see Figure 14).

---

[18] Bellcore TA-TSY-000310 Issue 2, March 1987.

[19] Almquist, M.l and Bowyer, L.R. GLOBECOM '86: IEEE Global Telecommunications Conference (Cat. No. 86CH2298-9) Houston USA 1-4 dec. 1986 Vol.3.

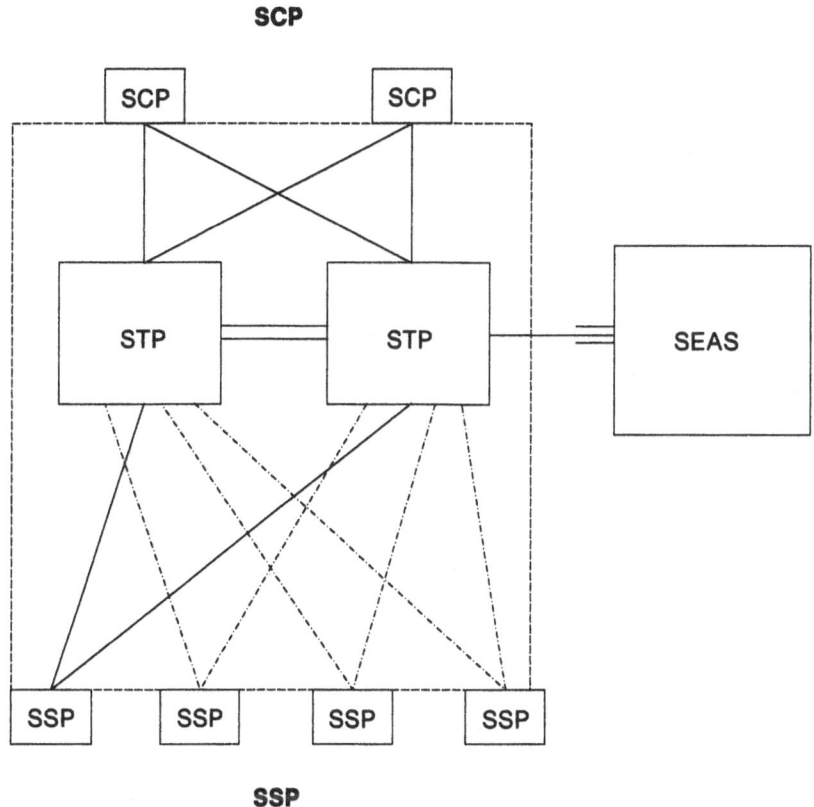

Figure 14. SEAS Control of CCS7 Network

Network-element-specific management functions are usually executed via an operator console. This console is located, for example, inside the SSP, but the functions may also be executed from a remote location via the operator consoles of an operations and maintenance center that may be identical to a SEAC.

Configuration Management
Configuration, in relation to the CCS7 network, includes:
- Setting up a CCS7 network, for example, by adding new signaling points (SSP, STP, and SCP).
- Changing the existing CCS7 network, for example, by adding signaling links between existing signaling points.

The tools for configuration management actions will be network-element-specific and implementation dependent. (The functions listed in "Monitor and Control" will be used to execute the steps in setting up or changing a signaling network. Measurements relating to links must be made at both ends, but not necessarily at the same time.)

CCITT defines additional functions for the Network Manager "Monitoring and Measurements for the MTP", specified in Q.791. About 60 measurements are reported for a defined time-interval, for example:

- The number of MSU, for example, per signaling link
- The octets, for example, per signaling link
- The events (such as all the functions in case of failures or overload mentioned in "Monitor and Control")
- The duration, for example, of unavailability or overload.

Many of these measurements are not compulsory and thus depend on national specifications. When completed, Q.795, "Operations and Maintenance Application Part" will specify the control of the above measurements and the handling of data collected.

The SEAS supports the Network Manager with a set of traffic load, utilization, and performance data. These measurements partly overlap those defined by CCITT.

In later releases, more sophisticated engineering techniques are likely to be employed.

SEAS contains a database of the current and planned CCS7 network configuration[20].

Monitor and Control

CCITT standards and Bellcore requirements define the following automatic functions to keep a CCS7 network running:

- Bringing links into operation or stopping operation (Q.703)
- Detection of failures (Q.703)
- Directing traffic via alternative links or routes (Q.704)
- Redirecting the traffic to links synchronized again or physically repaired (Q.703, Q.704)
- Detection of overload situations, and reduction of traffic at its source (Q.703, Q.704, 1TR7, Bellcore).

Which alarms are given in a failure situation are determined by national specifications, for example 1TR7 of Deutsche Bundespost (DBP).

The SEAS surveillance and administration functions provide continuous performance monitoring of STPs and signaling links, by watching thresholds and exception reports of traffic and performance measurement data items. This data may overlap with the network engineering data. (See "Configuration Management". It gives a detailed overview of the situation in the CCS7-node parts of the SCP and the SSP; see Figure 14).

---

[20] Almquist, M.l and Bowyer, L.R. GLOBECOM '86: IEEE Global Telecommunications Conference (Cat. No. 86CH2298-9) Houston USA 1-4 dec. 1986 Vol.3.

Maintenance

The functions listed in "Monitor and Control" can also be used to take parts of CCS7-nodes out of operation for repair and bring them back to operation. Physical repairs are implementation dependent and must be made at the node location.

New software releases will be introduced into a signaling network step-by-step. Therefore, a new release must at least be compatible with the previous release running on the adjacent signaling nodes.

Features based on new software releases can be put into service after the complete software change in the relating network, and then started via the craft terminal.

The EWSD-SSP will facilitate the introduction of new software releases without interrupting established calls.

Administration

The routing data in the STP, SSP, and SCP and the data for global title translation in the STP must be administered in a CCS7 network.

The data on the respective network elements is provided via the craft terminal (MML).

SEAS provides the MML capability for the STP by its "Recent Change" functions. Standard masks aid the user in the manual entry of STP translation data.

Testing

CCITT has defined the following testing functions:
- The data link is tested automatically during the initialization phase (Q.703).
- The signaling link test (Q.707) ensures the communication between both ends of a link.
- CCITT defines in Q.795 functions to verify routing data.

SEAS can verify the translation data in the STP.

3.6 IN-Services and the SMS

This section is a functional description of the Service Management System (SMS), which is largely based on the version of the SMS used to support the RBOC 800 service. The following general requirements have also been considered:
- The SMS should support a number of service offerings, in addition to the 800 service.
- The SMS should function in countries other than the U.S.A (this means that it should be adaptable to national X.25 networks, and accommodate different national languages and operating conventions).

3.6.1 Service Management System Overview

A Service Management System (SMS) is the operations system through which
network operator and service subscriber personnel manage SCPs and related service
applications (programs and databases) in an IN (see Figure 15).

More than one SMS may be associated with the IN; the network operating
company may want a separate SMS for each IN service or a single SMS for several
IN services.

The SMS resides in a multipurpose computer. Processing power and database
size requirements normally govern the choice of a specific computer. The SMS
manages a private network consisting of switched and leased lines connected to a set
of keyboard or display terminals through which network operator and service
subscriber personnel gain interactive access to the system.

Access to the set of SCPs managed by the SMS goes through an X.25 network.
The number of links from an SMS to the X.25 network will depend on how much
traffic must be supported. In addition, the SMS has a communication link or
magnetic tape interface to the SEAS and may be connected to a network
operator-provided billing system.

If a single SMS is considered to be insufficient, a backup SMS should be
provided. It should be constantly maintained, so it can rapidly assume control in
the event of the main SMS failing.

A number of user classes are associated with an SMS:

- Service Operator personnel, responsible for installing and maintaining a service
  offering
- Service subscribers from private companies
- Network operator personnel, responsible for installing and maintaining the
  SMS
- Software development personnel, from various organizations responsible for
  SMS and service program maintenance.

Network Operator/Service Subscriber Terminals

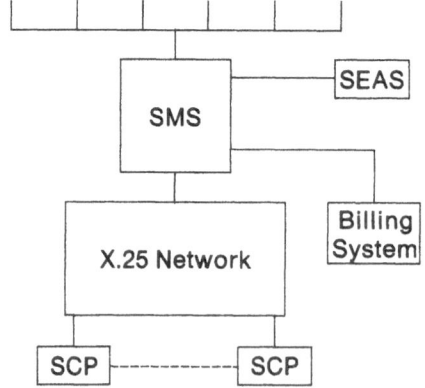

Figure 15. SMS Overview

The terminal interface presented to service operator personnel and service subscribers will be efficient even for inexperienced end-users. SMS software is structured so that the interface can be tailored to national languages and usage conventions.

The SMS base software is structured to provide a basic platform on which programs supporting individual service offerings can be built. The functions provided by the SMS can be classified as:

- Functions common to all service offerings
- Functions specific to an individual service
- Functions used by professional programmers and computer operators to install, operate, and maintain the SMS.

(These functions are common to all complex databases and data communication systems and are not considered here.)

3.6.2 SMS Function Description

User Access Control

This function is common to all service offerings and must be used by all service subscribers and service operator personnel who access the SMS.

Dial-up or dial-back control is provided on switched lines.

Each user is assigned a user ID and password which he can change.

A profile is assigned for each user identifying what functions he may use, what data he may access, and the type of access allowed. (For example, the Green Number Service provider requires read access to all assigned green numbers for summary reporting purposes, whereas the owner of a particular green number requires change access to the routing or scheduling information underlying his particular number.)

An user logging function monitors which users logged-on, and when, and what particular data modification functions they have executed during their log-on period.

Database Control

The SMS supports one or more databases corresponding to the services it administers.

Each database is accessible from service applications using a set of functions provided by relational database management systems. Classical database back-up, recovery, audit and integrity control facilities are provided.

Each service subscriber may have full read and write access to the entries in the databases which define his personnel data. He may update his data and provide scheduling information to indicate when his changes are to become effective. A user may employ delayed scheduling which means that the update should become effective at some future time and date specified by the subscriber, or immediate scheduling where the update should become effective as soon as possible.

The service application verifies input, and the appropriate feedback (error messages or update confirmation) is sent to the subscriber.

Each service operator has full access to the database corresponding to the service for which he is responsible.

The SMS automatically provides common services for all databases and all users. In particular, it will:

- Manage the download of initial versions of the databases to the appropriate SCPs. Very large databases will use magnetic tape when downloading, otherwise downloading will take place automatically over an X.25 communication link.
- Ensure the prompt update of databases on the back-up SMS, where appropriate.
- Keep centralized control of all scheduling information, in order to trigger the SCP database updating at the right time.
- Ensure the correct transmission of database changes to the appropriate SCPs.

SCP Traffic and Performance Management

The SMS automatically receives traffic and performance data from the SCPs it controls. This data, which generally takes the form of a set of counters produced by the components of each SCP, can be classified as:

- Signaling System Number 7 related counters.
- Individual service traffic and performance data.
- SCP implementation-dependent traffic and performance data. (This class of data will directly reflect the implementation in terms of the specific organization of processors and of their interconnecting links and the organization of software within this hardware framework.)

These classes of data are stored and managed by the SMS. Subsets of the counters are automatically transmitted (or sent using magnetic tape) to the SEAS, for use in CCS7 network management.

The SMS has a number of programs to process the data:

- Automatic analysis programs to detect emergency situations, for example, overload, and to report these to online SMS operators.
- Programs to respond to interrogation requests from network and service operator personnel.
- Background analysis and report generation programs.

Service Maintenance

Service operator personnel must examine and resolve the complaints of individual service subscribers. Thus, as well as providing access to the SMS database, functions are provided whereby the appropriate SCPs can be interrogated and controlled from the SMS. In particular:

- Dynamic data may be requested for a specific destination (for example, calls to a specific Green Number)
- Dynamic data may be requested for a specific origin (for example, calls from a specific subscriber number)
- Current SCP status of a given service may be requested (for example, usage and availability data).

Underlying these specific service-related SMS functions are:

- A common SMS conversation capability allowing any service application in the SMS to communicate on-line with its counterpart in the SCP.
- Appropriate support functions in the individual service applications, resident in the SCP nodes.

SCP Node and Service Support
The SMS manages sets of configuration data corresponding to the SCP nodes and services under its control. This data can be classified as:

- Signaling System Number 7 SCP node-related information (where this is not already supplied by the SEAS). This type of data is entered by network operator personnel and transmitted to the appropriate SCP on demand.
- SCP node implementation-specific data (such as disk or buffer allocation parameters). This type of data is also entered and sent to the SCPs by network operator personnel.
- Service specific control data (such as queue sizes, timer values). This type of data is entered by authorized service operator personnel on a service by service basis, and transmitted to the appropriate SCP on demand.

SCP node and service support functions are normally used:

- When the SCP node or an individual service is installed.
- When the SCP node or the services must be changed to cater for traffic growth or emergency situations.
- To balance resources among the various services.

Billing
Two levels of billing function are provided:

- Service subscriber is billed for his specific usage of the SMS.
- Service provider is billed for the SMS activity on the overall service which he provides.

Charging rates depend on factors like log-on duration, database access, and the reports generated.

If the SMS is connected to an external billing system, each service is responsible for generating billing records in a specific standard format. Where necessary, the SMS is responsible for intermediate buffering of billing records and for their reliable transmission to the billing system.

Report Generation
The SMS provides a set of report generation programs, concerning activity in the SMS and the various SCP nodes which the SMS administers. Reports concerning overall node and individual service activity can be generated, including:

- Traffic and performance analysis
- Billing information
- Terminal user activity
- Availability statistics.

3.7 Administrative System Functions and IN Requirements

In the USA (and probably elsewhere) the three management systems serving the basic network layers have been designed and implemented as three independently-operating administrative packages:

- An OSS, which manages one or more SSPs of the same type.
- A SEAS, which manages a set of interconnected STPs belonging to an CCS7 network.
- An SMS, which manages SCPs.

Thus, from an administrative view, an IN is viewed as a collection of unrelated physical and logical components.

The network operator's objective is to increase revenue and improve the quality of a service by providing an IN with a competitive set of enhanced services to service users and service subscribers. His primary goal, therefore, is to install, maintain and upgrade his services efficiently, and on an end-to-end basis. This objective is hampered by non-integrated management systems.

The IN administrative systems should provide the network operator with the tools required to meet his customer related goals.

When IN services are introduced into networks, the existing management systems are initially able to provide IN service management. However, as the amount of services and/or service usage grows, a more sophisticated and comprehensive service management must be developed in order to provide the service user and service subscriber with a grade of service which will stimulate service growth.

3.7.1 IN-Wide Service Management Examples

Service Subscriber Complaint Handling
Service subscribers require fast responses to their service problems and expect corrective actions to be taken promptly.

Service maintenance facilities located within an SMS and only having access to a set of SCPs are only useful if the underlying problem is located in one of these components. If the problem lies elsewhere (for example, in a SSP or somewhere in the CCS7 network), other administrative functions will be called on. Currently (via the OSS and SEAS) there are no service maintenance facilities at their disposal, so that problem diagnosis could be long and inefficient.

The following are some examples of how an IN-wide NSMS could improve problem diagnosis.

This situation could be helped by an IN-wide approach, consisting of a service-related trace facility in which:

- Each tracing request is activated or deactivated from a single terminal by service operator personnel
- Each tracing request specifies what service is to be traced and gives the appropriate service subscriber coordinates

- Activation or deactivation of a tracing request is automatically transmitted to each IN element.
- Tracing output, if any, is returned to the user terminal for presentation automatically and in an easily understandable format.

Service Application Program Installation and Updating

Service application programs must be installed and tested in the SMS, SCP, and SSP. New releases must be installed and become operative with minimum disruption to existing service subscribers.

In the current element-oriented approach, initial installation and subsequent updating of a service application program is a complex process that must be done manually. There is no interconnection between the SMS - SCP element domain and the OSS - SSP element domain.

In the IN-wide approach, the loading of programs into individual elements would be element-dependent, whereas the coordination of program activation is automatically driven from a central point.

Traffic and Performance Analysis

Each element in the IN can measure the use of its local resources. This information is periodically collected and stored by the corresponding administrative system. The system uses the information to perform traffic and performance analysis at the element level (SMS and OSSs) or the sub-network level (SEAS). Currently, there is no facility for IN-wide traffic and performance analysis, which means that analysis of the end-to-end performance of the network services is difficult.

A centralized focal point responsible for IN-wide traffic and performance analysis is needed. Each administrative system could be connected to the focal point and could periodically send a subset of its local usage information.

3.8 Trends

Trends in network operations are important considerations in connection with the evolution of the networks themselves towards ISDN and the advent of a market-driven provision of network services.

3.8.1 Evolving Operations Architecture

The critical element in the evolution to an integrated NSMS is a practical method of migrating from the existing base of network management systems, while maintaining network integrity and efficiency for basic services. This migration must be based on a set of standards for management functions and interfaces.

OSS/NE Standard Interface

OSS/NE standard interfaces play a key role in the management and operation of the NEs from OSS applications and user workstations. Due to the lack of protocol standards, existing public networks have not been particularly suitable for the transfer and switching of data between the NEs and the OSS centers. Dedicated overlay networks have been introduced with supplier-specific formats and protocols. The standardization effort is directed towards communication between the NEs and OSS in the areas of language, structure and presentation of OA&M messages and bulk data.

The interface definition includes protocols, messages and formats. To ensure compatibility Operations Systems Modifications for the Integration of Network Elements (OSMINE) is defined in the US. It is a new process, in which product suppliers can participate, providing compatibility analysis and OSS/NE testing.

OA&M Network Management Integration

One solution to the problem of how to integrate the existing operation systems is to replace the dedicated operations systems network with the public network itself:

- Using exchanges as data switches
- Using the CCS7 network as a bearer for OA&M messages
- Connecting operations terminals via ISDN access.

Human Interface

Today the operator communicates with the switching system using a non-intelligent dialog device (for example, a display terminal and printer) with the aid of MML commands and messages (mostly based on CCITT recommendations Z.3NN). The MML is usually overlaid by supplier specific menus and formats. One example is the extended MML of the EWSD. Evolution in this area is concerned with:

- Use of improved input and display methods using PCs, graphics, pull down menus, mice, icons, and so on, as in office environments
- Uniform operating methods at remote and local operations terminals
- Uniform customer interface, thus integrating the various OA&M user groups
- Security and authorization
- System guidance replacing operating manuals and reducing human errors. Expert systems could be used for fault management.

Future Network Management Systems Architecture

Standard interfaces for the connection of real open systems and networks must be defined as a framework for interconnection and a set of protocols and formats that support the transfer of information for application processing. The Open Systems Interconnection (OSI) reference model is a commonly accepted international standard for this task and the main protocol design recommended by CCITT and ISO. A network management systems architecture must cover:

- Classes of applications
  - Transactions
  - File transfer
- Upper layer protocols

- Data network access, for example X.25, CCS7/MTP/SCCP, ISDN Q.931, and HDLC LAPB/LAPD by using dedicated circuits.

Figure 16 shows the protocol architecture of an OA&M network being studied by CCITT.

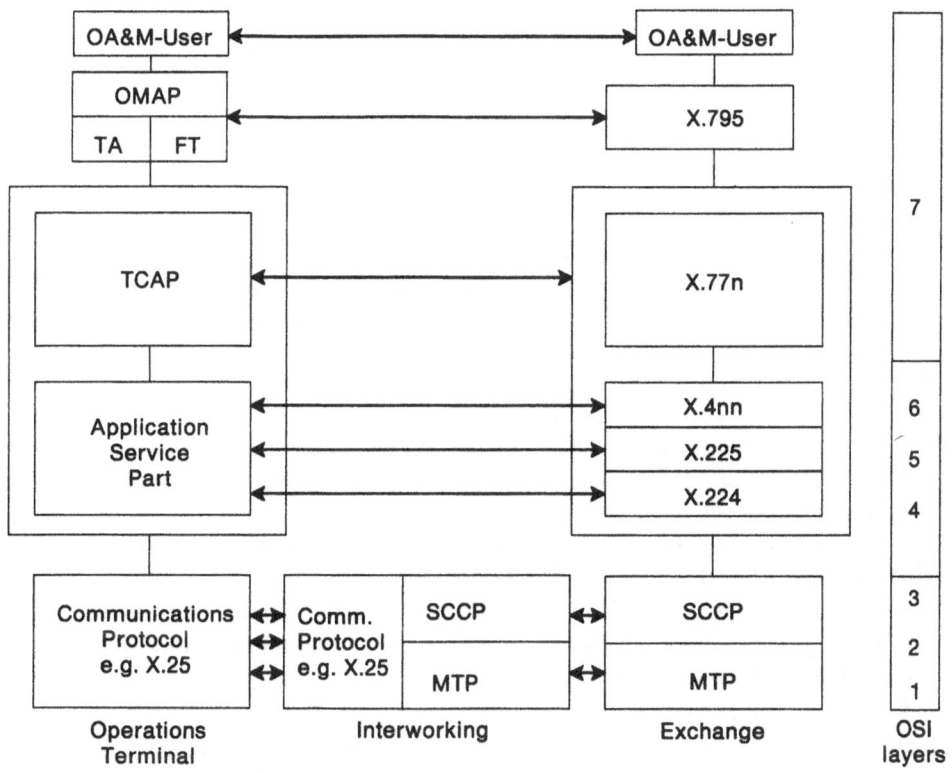

TA = Transaction
FT = File Transfer

Figure 16. Protocol Architecture in OA&M Networks

Chapter 4. Network Components

4.1 Main Objectives of the IN Architecture

Telcos must become more and more market-focused due to the evolving multi-vendor/multi-service provider scenario all over the world.

Until very recently, the process of introducing new (telephone) services to customers was a long and costly one. Before a new service could be introduced, Telcos were forced to rely on switching system vendors to enhance the capabilities of their systems. Also, before a Telco could make the new services generally available, switches had to be upgraded throughout the network. This situation will change with the advent of the IN architecture.

From a network operator's point of view, the main objectives and thus the main engineering requirements for the IN architecture are:
* To reduce time and cost of service creation, customization and introduction
* Without jeopardizing with the network's integrity, reliability and availability
* With minimum performance losses
* And saving the existing network investment; for example, changes to the existing trunk signaling configuration for the IN cannot be assumed.

4.2 Network Topology

This chapter examines the IN network topology and describes the components and interfaces for service delivery. As the architecture must built upon the existing network, the IN core network (consisting of SSPs, STPs and SCPs) and the peripherals are distinguished. The composition of individual IN components is such that:
* The SSP can be provided by standard EWSD hardware and software products (such as local or tandem exchange equipped with CCS7) augmented with the following software subsystems:
    - SSP triggers
    - SSP application part
    - SCCP
    - TCAP.

- The STP can be provided by standard EWSD hardware and software products in an overlay network architecture augmented minimally by an SCCP (class 0 and 1).
- The SCP can be provided by standard EDP hardware and software products (such as IBM/370, PS/2 and software program products or, for instance, Siemens 7.5xx, TRANSDATA front ends and Siemens EDP software program products, each configuration serving alternately as a second source to a Telco) augmented by the following subsystems:
  - CCS7 functions hardware and software
  - SCCP software
  - TCAP software
  - Node Manager software
  - SMS/OSSs hardware and software interface
  - Application platform software
  - Application software.
- The SMS is only generally defined as an administrative system reached via an Operations Systems Network Interface (OSNI). The SMS components and the application content will probably be country-specific.

4.2.1 Performance and Throughput Requirements

Performance

The response time for a service user is the period of time from the completion of dialing until the receipt of audible ringing, and is referred to as Call Setup Time (CST).

As it has a greater capacity compared with other signaling systems, the CCS7 network reduces this response time, but the introduction of an IN architecture tends to increase the CST due to the following:

- Forwarding the virtual number to an SSP in a C2-environment (see 1.4, "IN Introduction Scenarios") where the CCS7 network is only available at the tandem level.
- The SSP detecting a special call (the virtual number), interrogating a trigger table, and initiating a service query to get additional data for call completion. The period from when the SSP sends a query to the SCP until it gets an answer should be less than three seconds. Where a dialog (of messages) is necessary between the SSP and the SCP, response time covers the time for one dialog step.
- The CCS7 network routing the query to the appropriate SCP and the response back to the SSP using one or more STPs, depending on the CCS7 network topology. The permissible delay time in an STP is 20 to 100 milliseconds, depending on message length and whether or not global title translation is performed at the SSP level. In case of failure, the delay time is between 100 and 400 milliseconds. The transmission delay within the CCS7 network is negligible.

- For the SCP interfacing the CCS7 network, interrogating the service database and delivering the response message to the CCS7 link, the response time for a single step retrieval transaction (for example, a provide information request for GNS) should not exceed 0.5 seconds (average), and 99% of such transactions should not exceed 1.0 second from receipt of the inbound message at the signaling point interface until the response is delivered to the CCS7 link.

The response time for a service subscriber relates to his interface to the SMS and to the SMS-SCP interface.

- The subscriber's interface to the SMS is a dialog interface with the same response time requirements as other management and information systems, such a general response time for minor transactions, like changing a menu or requesting help information, should average one second and in 95% of cases be under three seconds. The response time for updating service information depends on the complexity of the update and should be within 15 seconds, or must be done asynchronous or in batch.
- The response time of an SCP receiving an SMS message should be eight seconds (average) and in 99% of cases should not exceed 16 seconds.
- There should be an express function for downloading service updates from the SMS to the SCP. The delay between the input by the service subscriber and the availability of the update in the SCP database should be 15 minutes (or shorter for special services like PVN).

Throughput

- The expected traffic volumes generated from the selected services can be handled by the product solutions for the SSP, STP, and SCP described later.
- An SCP must be able to serve more than one service.
- The flexibility of the described product solutions means a wide range of database sizes and transaction volumes can be handled.
- An SCP node must have the built-in ability to be upgraded without disruption to service.
- The architecture must be able to handle a variable number of SSPs, STPs and SCPs.

4.2.2 Availability Requirements

Availability requirements depend on service quality standards of the different network operators. The following requirements are based on the Bellcore reliability and quality requirements for switching systems, including the SCP[21].

- SSP and SCP (node portion, signaling point)
  These components should be available 24 hours a day, all year, with a maximum downtime of three minutes a year.

---

[21] LSSGR LATA Switching Systems Generic Requirements, Bellcore TR-TSY-000064.

- CCS7
  The downtime from one signaling point to another should not be greater than 10 minutes per year, including the transmission path and the signaling endpoints.
- SCP application
  The downtime for one SCP application should not be greater than 10 to 20 hours per year for one site. The downtime for a service from a mated pair should not be greater than three minutes per year.
- SMS
  The SMS should be available seven days a week for 22 hours a day (everyday 2 hours are used for SMS maintenance).

4.2.3 IN Architecture

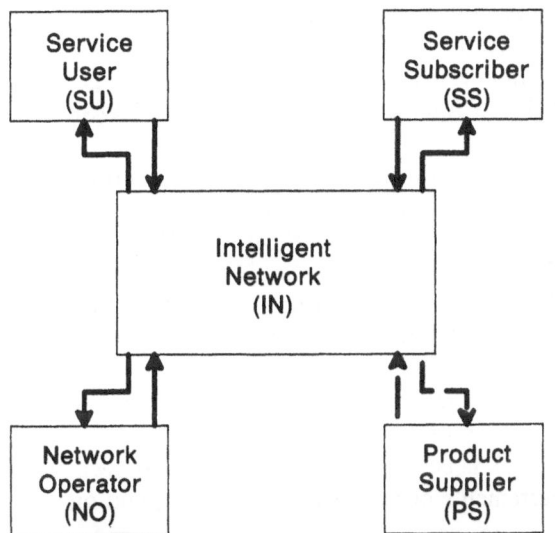

SU-IN Access code (service) and virtual number (service subscriber). If the service includes any form of prompting between service and user, POTS without MF signaling cannot be used.

IN-SU Audible ringing and call setup. With enhanced services, the user may be prompted via announcements to enter additional digits.

SS-IN Service requests and service related data (for example, the real number, routing and screening parameters, alternate billing number).

IN-SS Request confirmation, data update confirmation, dialog guidance and help information, billing information and service statistics.

NO-IN Supply, operations support, administration and maintenance of service data and logic. The service network operator may be a PTT or a third party. The PTT remains responsible for operations support, administration and maintenance of the underlying network if the service network operator is a private organization.

IN-NO Network operations data, traffic measurements data, billing information.

PS-IN IN network resource provisioning and assistance to the network operator within the area of operations and maintenance.

IN-PS IN product requirements.

Figure 17. IN User Interfaces

IN Network User Interface

Figure 17 shows the user interfaces to the IN. The network itself is seen as a black box to the user, regardless of the specific network component to which a user interfaces.

IN Core Network and Peripheral

The IN architecture provides network connection control intelligence at centralized nodes (SCPs) to switch connections under the direction of the SCPs within the existing network. In this chapter, the IN core network and its peripherals are distinguished.

The core network consists of components which must be engaged in call setup, such as the SSP, STP, and SCP. Peripheral components access the network, support call setup procedure, prepare additional service logic and/or data, or support, administer and maintain the networks service delivery.

Figure 18 shows the components and their related interfaces.

IN Core Network

The core network consists of SSPs, STPs and SCPs connected by CCS7 links.

The SSPs are switches with CCS7 facilities (MTP, SCCP) and detect special service calls, and interrogate a remote database in the SCP via TCAP messages. The SSP is the signaling endpoint of the CCS network, and the Siemens EWSD is the proposed solution.

The STPs are Signaling Transfer Points with the ability to route signaling messages between signaling endpoints, for example to route service queries from an SSP to the appropriate SCP. The proposed solution is the Siemens EWSD.

The SCP node, acting as a signaling end-point within the CCS network, can handle the service queries from the SSP, to deliver the service data to the SSP and to process an operational database. The SCP is built using general purpose computers; the described solution is IBM computers.

The SCP, STP, and SSP must be able to interface with network components from other vendors (multi-vendor, second sources to the Telco).

The configuration of the core network is country or Telco-specific and depends on the reliability and availability requirements of the network operator, and on traffic requirements.

- CCITT recommendations provide several options for the CCS7 network, therefore the network operator can use an associated CCS network configuration (where the endpoints of a user trunk group are interlinked directly by means of one or more CCS links) or a quasi-associated CCS network (with stand-alone STPs as an overlay network). A third alternative is to use a combination of an associated and a quasi-associated CCS network.
- The network operator can use the SCPs that support more than one service or a service can be located on more than one SCP because of security requirements or load sharing considerations.

Figure 19 depicts an example basic data flow between an SSP and an SCP for a GNS without a prompting option.

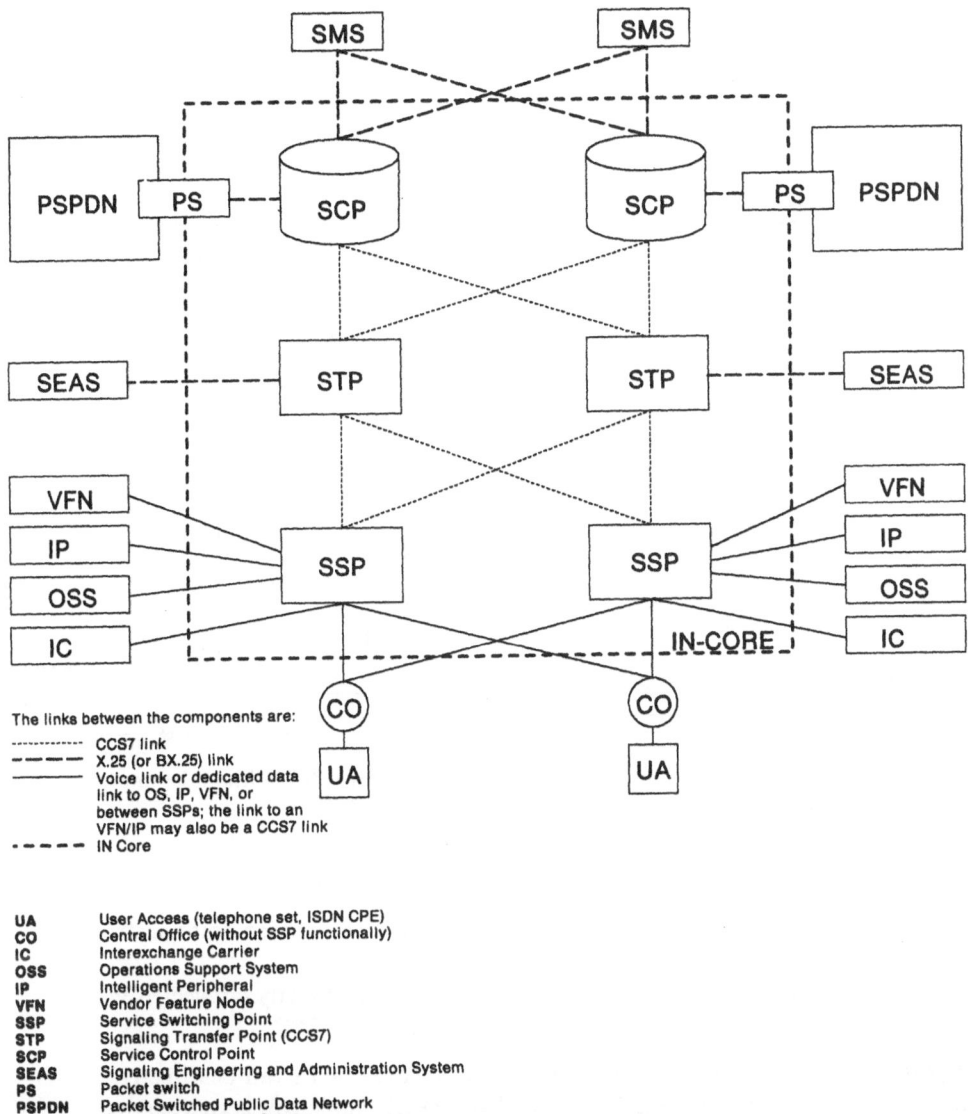

The links between the components are:

---------- CCS7 link
– – – – X.25 (or BX.25) link
————— Voice link or dedicated data
link to OS, IP, VFN, or
between SSPs; the link to an
VFN/IP may also be a CCS7 link
- - - - - IN Core

UA	User Access (telephone set, ISDN CPE)
CO	Central Office (without SSP functionally)
IC	Interexchange Carrier
OSS	Operations Support System
IP	Intelligent Peripheral
VFN	Vendor Feature Node
SSP	Service Switching Point
STP	Signaling Transfer Point (CCS7)
SCP	Service Control Point
SEAS	Signaling Engineering and Administration System
PS	Packet switch
PSPDN	Packet Switched Public Data Network
SMS	Service Management System

Figure 18. IN Components and their Related Interfaces

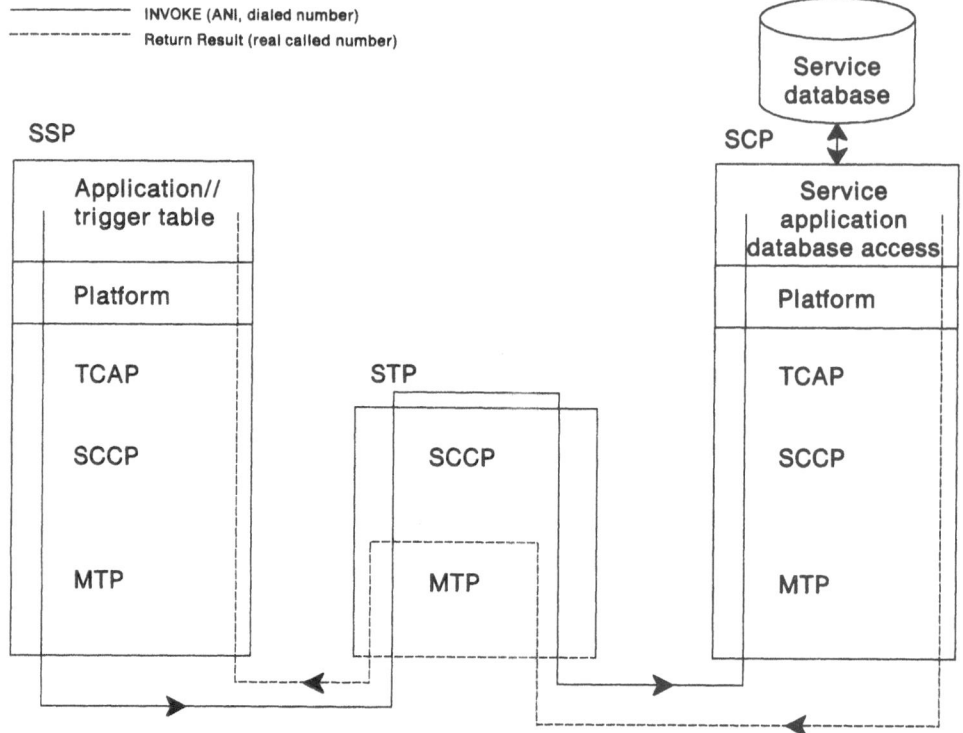

Figure 19. Basic Data Flow Between an SSP and SCP (for GNS Without Prompting
Option)

Peripheral Network Components
User Access and Local Exchange: It is assumed that the service user's instrument is
a telephone set or ISDN CPE.

The local exchange is a switch to which the user is directly connected via an
existing user access loop. It may have the facilities of an SSP.

Packet Switch: According to Bellcore requirements[22] SCPs can be directly
connected to PSPDNs by X.25 links so that PSPDN node can access service
applications resident on the SCPs. An SCP should be able to accommodate a
number of links to packet switches within a PSPDN.

A PSPDN switching component may act as an SSP. The SSP begins a
service-related query, triggered by an external stimuli, which is routed to the SCP
via PSs.

---

[22] Service Control Point Node—Generic Requirements, Bellcore TA-TSY-000029.

Vendor Feature Node: A VFN may be connected to the IN network by voice or data link, or by a CCS7 link.

Intelligent Peripheral: See on page 11 in 1.2.1, "IN Elements".

Signaling Engineering and Administration System (SEAS): The SEAS is the operations support system for the CCS7 network.

Operations Support System (OSS): The operations support systems allow the network operators to monitor, control, administer and maintain the network elements.

Service Management System (SMS): The Service Management System provides administrative support for the SCP node and SCP service applications, such as providing data to the SCP and collecting traffic and performance measurements from the SCP.

4.3 Service Control Point (SCP)

4.3.1 Requirements

The SCP architecture and functions are in accordance with CCITT and ANSI recommendations for Signaling System 7, and the Bellcore documents:
* *Service Control Point Node Generic Requirements*, Bellcore TA-TSY-000029
* *Bellcore Specifications of Signaling System 7*, Bellcore TR-NPL-000246
* *SCP Node/SMS Generic Interface Specification*, Bellcore TA-TSY-000365.

Node Definition
The SCP is a real time, high availability system hosting database applications which provide enhanced network services. Within the IN, the SCP is an addressable entity or node in a Common Channel Signaling Network (CCSN) and/or a Packet Switched Public Data Network (PSPDN). It provides a network interface to resident service-providing applications which generates responses to service-related queries received from other nodes in these networks.

The SCP provides all functions to support one or more IN applications. An SCP site comprises the hardware and node software necessary to support these applications. Node functions are independent of, or common to, all applications.

Sufficient processing and storage capacity must be provided to satisfy the high performance requirements of an SCP site. Additionally, the configuration must accommodate growth without disruption to service.

SCP Processing Functions
The SCP's primary task is to provide highly reliable access to, and processing of, database applications. The SCP must allocate its resources to perform the following functions:

Message Processing: This comprises those functions required to deliver a message to its proper destination within the SCP or to the appropriate node on an attached network. These functions include:

* Execution of network interface protocols (CCS7, X.25, BX.25)
* Protocol error handling
* Message discrimination
* Message distribution and routing
* Network management and testing
* Application processing
* Node administration.

All service-related messages (application queries and responses), network management messages, and administrative messages associated with support systems are handled by the message processing function.

The SCP must be able to interface with many networks simultaneously to process service-related, and operations and administrative messages.

Node Operations, Administration and Maintenance: These functions are required to operate and control the SCP itself:

* Performance measurement collection and reporting
* Security
* Error and fault detection, notification, isolation and recovery
* Overload detection and flow control
* Health and status monitoring
* Configuration control.

SCP External Interfaces

The IN model defines two sets of interfaces for the SCP: service network interfaces and support system interfaces, both of which are required to provide database application services.

Service Network Interfaces: The SCP interfaces with a CCSN and/or PSPDN to process service-related messages. The interface must present high availability to the networks as the CCSN requires high availability at each signaling end point. Whether an SCP interfaces to both a CCSN and a PSPDN depends on the requirements of the resident application(s).

The SCP interfaces with a CCSN using the CCS7 protocol. A front-end communication function of the SCP provides the link termination for the transmission link which physically connects the SCP to the CCSN. The CCSN end of this link generally terminates at an STP (or any other adjacent signaling point). The number of these links is determined by the volume of traffic presented to the SCP.

The SCP may communicate with a PSPDN using the physical, link, and packet levels of the X.25 protocol. Transmission links connect the front end of the SCP with a packet switch providing entry to the PSPDN.

Support System Interfaces: In addition to the service network interfaces, the IN architecture will define SCP interfaces to Operations Support Systems. These are:

* Service Management System (SMS)
* Signaling, Engineering, and Administration System (SEAS)
* Remote maintenance operations centers
* Local maintenance operations centers.

Transport between these three support systems and the SCP is provided by the Operations System Network (OSN), with the SCP accessing the support systems via a generic OSN interface. However, for the purpose of introducing these support systems, they may be considered to have distinct interfaces to the SCP.

The SCP employs the physical, link, and packet layers of the X.25 protocol for the transmitting and receiving of messages to and from an SMS. The SMS provides administrative support for the applications resident at the SCP. SMS administration of the resident applications entails maintenance of the application databases, and collection of traffic and performance measurements from the SCP. Sufficient transmission capability must be provided between an SCP and SMS so that each application hosted by the SCP is allocated a virtual channel to the SMS.

The SEAS collects performance measurements regarding CCSN components (including the SCP) for network engineering and administration.

The SCP also communicates with the maintenance operations centers using the X.25 protocol. The maintenance operations center receives alarm and status information from the SCP so that the SCP may be monitored remotely. The maintenance operations centers can also control SCP operations (including system reconfiguration, initialization, and emergency actions) in order to meet SCP reliability, availability, and performance requirements.

Service Applications

Each application contains the logic and data necessary to provide specific network services. An application may respond to a query by providing call completion instructions or information supplemental to call processing. Furthermore, a given application may provide more than one service with each service using the common base of data contained in the application's database.

It is important to note that the processing and storage demands made on the SCP may vary significantly from one application to another.

An SCP supporting more than one application must be capable of appropriately allocating its resources to satisfy the differing storage and processing requirements of the resident applications.

4.3.2 SCP Architecture

The Service Control Point (SCP) provides high-volume database transaction services in a realtime, multi-tasking, high-availability environment.

SCP requirements are met by the system design through a combination of modularity and redundancy. Discrete components are loosely coupled in a unified manner to maximize throughput and performance.

For extremely high network availability, SCPs or STPs may be deployed in geographically separated mated pairs. It may be desirable to collocate the paired SCPs with the paired STPs on which they are homed.

One approach is to outline a conceptual IBM architecture that would attach several System/370-based application host processors to a mated set of token ring LANs. With this approach, connections between token rings and application host processors would be supported by standard hardware and software products.

This SCP design would include elements of the IBM System Application Architecture (SAA), a collection of selected software interfaces, conventions, and protocols as a framework for developing consistent applications across IBM information processing environments.

For maximum flexibility, the SCP architecture is modular, comprised of the following major components (see Figure 20).

* Connectivity
* Front end
* Back end
* Node manager.

Connectivity
Connectivity provides reliable communication between the other three components using a dual token ring for greater speed, availability, capacity and flexibility.

Each of the three components consists of token ring-attached processor subsystems, providing appropriate capacity and availability. For each component, replicated processors are configured with like hardware and software for uniformity and interchangeability. Replicated processors are normally operated in a load balancing manner for optimum performance.

The dual token ring is a cornerstone of the proposed IBM SCP system design for adaptability and expandability of configurations.

A family of configurations conforming to a single set of system design attributes is envisioned for the IBM SCP to solve a wide variety of transaction rate needs and application requirements. Combinations and quantities of components are selected to optimize SCP functions and features, while providing configurations that can be supported.

The total number of SCP components is guided by the principle of N + 1 for reliability. Capacity requirements are based on the number of service networks, support network links, and peak busy second traffic load requirements. Traffic load capacity can be augmented by adding more application hosts or Signaling Point Interface (SPI) processors.

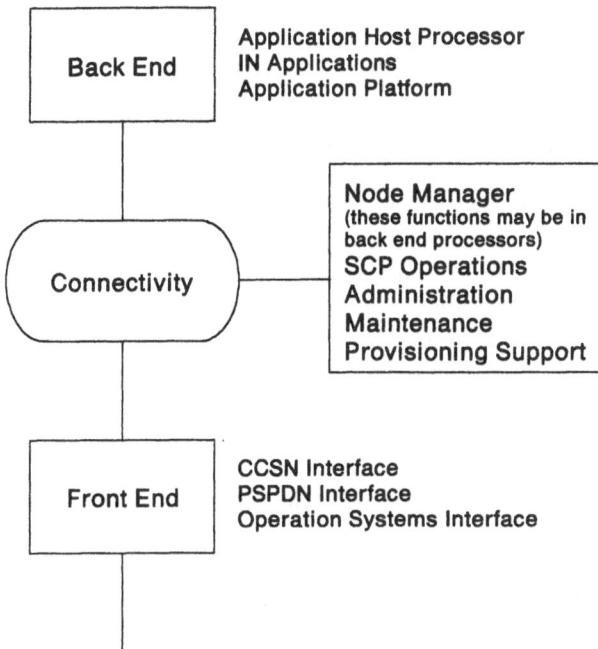

(To External Interfaces)

Figure 20. SCP Major Components

The number of SCP sites is determined by the applications selected and area traffic demography. Therefore, they must be individually engineered for each customer and each site. Availability characteristics for a given site must be engineered to meet those of the most stringent application in service at that site.

Front End

The **Front End** component provides the external communications interfaces to the service and support networks.

Each external interface is treated independently, so there is actually a series of front ends for an SCP. Each site can be engineered to provide only those interfaces relevant to the networks accessed from that site, while retaining the flexibility to add supported interfaces in the future. This allows interfaces to be tailored to network requirements. The external interfaces requirements are:

- The CCSN Interface, CCS7 (MTP levels 1-3 and SCCP)
- The PSPDN Interface, X.25
- The Operation Systems Network Interface, X.25 or BX.25 (Transport and UAL).

The CCSN and PSPDN interfaces are service network interfaces. The OSN interfaces are support network interfaces.

- Interface to Common Channel Signaling Network
  The signal point interface (SPI) provides the CCS7 interface between the SCP
  and the common channel signaling network (CCSN) for service-related message
  processing. The SPI receives service messages and signaling network
  management (SNM) messages from the CCSN and distributes them to the
  applications in the application host processors (AHP) or to the node manager.
  The SPI receives messages from the applications in the AHPs and from the
  node manager and routes the messages to the appropriate CCS7 link. Internal
  node message distribution is controlled by the node manager.
  The SPI participates in the CCS7 signaling network management and
  administration and maintenance procedures by collecting interface and node
  related measurements and communicating them to the node manager for
  forwarding to a management subsystem via the operations system network. The
  SPI software supports the ANSI or CCITT version of Signaling System Number
  7 for U.S. and international CCS, respectively (MTP and SCCP services).
  The total number of SPI processors is controlled by the capacity requirements
  of the SCP in terms of number of links required and amount of traffic to
  transfer. The minimum number of SPIs is two (given by the N + 1 principle for
  availability reasons). The maximum number is determined by internal node
  capacity constraints in the individual situation.
- Interface to Packet Switched Public Data Network
  Through a separate X.25 interface, the SCP can provide database services
  directly to a Packet Switched Public Data Network
- Interface to Operations Systems Network
  Operations Support Systems (OSS) distributed over the X.25 Operations System
  Network (OSN) access the SCP node and the SCP applications over X.25 links
  to the SCP. The Operations System Network Interface (OSNI) supports X.25
  links either over duplicated high speed data links or over packet-switching data
  networks. The OSNI handles the User Application Layer (UAL) and X.25
  processing for the SCP node. In addition, the OSNI handles the interface to the
  SCP Alarm Panel. The SCP Alarm Panel shows the status of the SCP and its
  components, and notifies the craft when the SCP status changes.

Back End

The **back end** component is a foundation for the database application services and
consists of System/370 processors hosting the applications and application
databases. The hosts are referred to as application host processors (AHP).

Each host processor contains platform node software up to, and including, an
application program interface (API). A given application may reside on two or
more hosts. An independent version of the application database is maintained by
each host processor on which the corresponding application resides. However, the
updates to an application database are synchronized across all hosts within an SCP
to provide a single system image to its external interfaces. An associated update
application is required if synchronization of several SCPs require synchronized
database updates.

The SCP back end is comprised of System/370 Application Host Processors (AHPs). The software environment includes standard IBM products to form the system software base, an application platform, and the specific application code, as shown in Figure 21.

- The Base System Software

  For growth and flexibility, the baseline or underlying software must consist of a fully supported mainline operating system, and associated program products. A wide range of systems are under evaluation for this purpose. State of the art communications functions are required; VTAM is being evaluated for the teleprocessing access method. The environment also needs a reliable and functionally mature transaction processing system. IBM has several advanced systems that are being assessed to perform this task. It is important that the major operating systems are optimized to work together to meet the specific requirements of an SCP node.

- The application platform

  The platform consists of common functions to be used by all IN applications, including data handling and control functions, communication functions, and application programming functions.

  Four major elements form the platform:

  - Data Communications, which distributes input to, and gathers output from, the applications.
  - Database Management, which processes data access and update requests, ensures data integrity, and handles data recovery.
  - Node Manager Interface, which processes all interaction with the node manager and performs back end processor administrative tasks on behalf of the node manager. It is a control point for the operation of the platform and the applications.
  - Application Program Interface (API), which allows telephone company operating-personnel to create new applications. This provides the potential for greatly reducing the time required to develop and implement new services into telephone networks.

Node Manager

The **node manager** component provides a centralized coordination point for the SCP node. Complete node management includes all coordination and control functions required to ensure continued operations, administration, and maintenance of the site. It includes the node manager component and node manager interface functions which reside on each processor subsystem of the front end and back end components. A primary objective of the node manager is to ensure that the SCP appears a single entity to its external interfaces. Functions supported by the Node Manager are:

- Service network management
- Support network management
- Local and remote craft interface
- Specification data and security administration
- Measurement collection

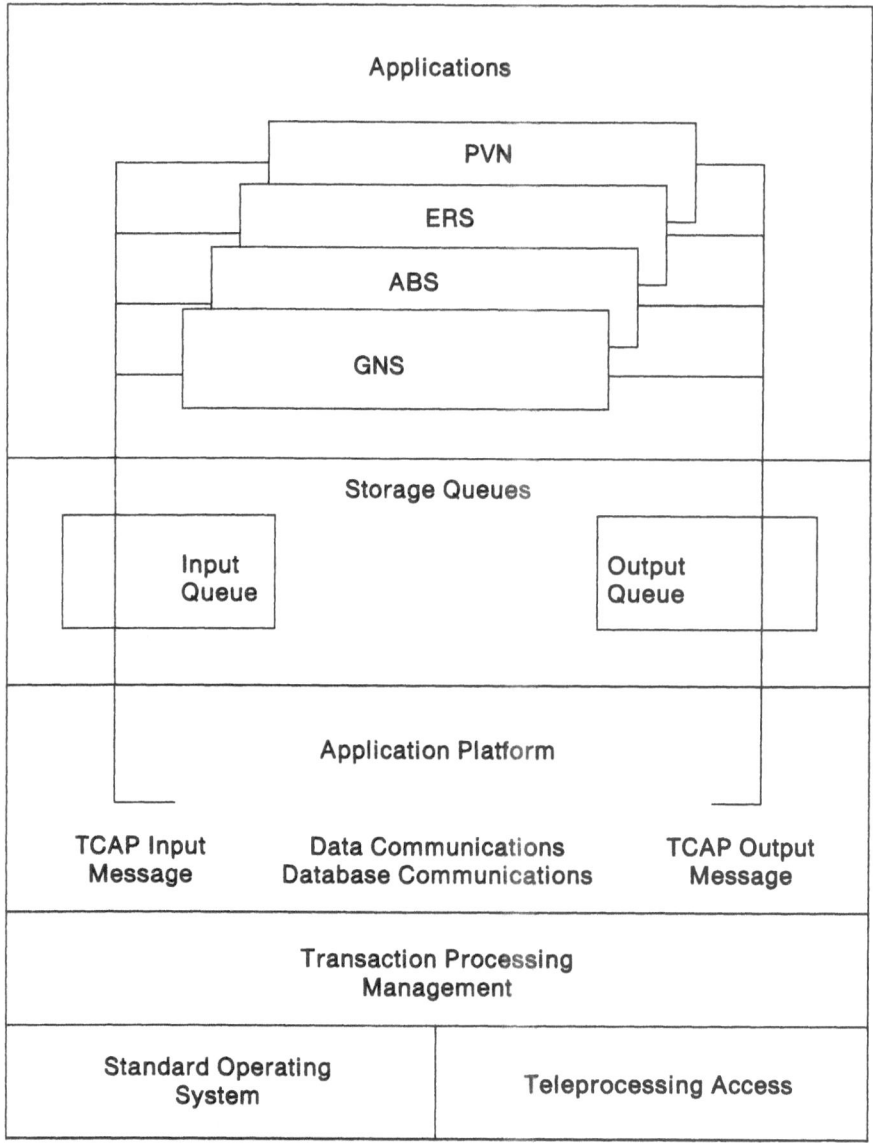

Figure 21. SCP Software Components

- Report generation
- Manual configuration and status control (hardware, software)
- Health and status monitoring
- Error handling and notification
- Diagnostics.

The node manager controls the SCP and provides a single system image of the SCP to the other nodes (STP, SMS, SEAS) and to the craft person. Node manager functions consist of performance management, maintenance, configuration control, administration and operational control.

- **Performance Management** consists of system status monitoring, measurement data collection, threshold analysis, load distribution, and MTP/SCCP network management.
  - System Status Monitoring:
    The node manager status monitor determines the operational status of the SCP from data obtained from each component.
    Each SCP component maintains internal status data that describes its current processing state and that of its active sub-components (such as the A-Links of an SPI). The internal status is sent to the node manager.
    SCP components periodically send a status message to the node manager indicating that they are operational. The period depends on the current importance of the SCP component and is supplied to each component by the node manager. Three levels of importance, critical, major, and minor, correspond to the three alarm levels. The period of each level is based on timing requirements of the corresponding alarm level. For example, an SCP component at the major level might report every 15 seconds (a major alarm may be raised in 30 seconds).
    Each SCP component is assigned a normal importance level which can be changed under certain circumstances. For example, the normal level of importance of an SPI may be major, but if only one SPI is operating, then its importance is raised to critical.
    If the node manager does not receive the status message at the expected time, it sends that SCP component a request for immediate response. If this request is not acknowledged, that SCP component is marked as inoperative, the other SCP components are notified to bypass it, and an alarm is raised. The node manager then tests the failing component to determine whether there is a problem. If there is no problem, the alarm is canceled, and all other SCP components are notified that the component is operational again.
  - Measurement Data Collection:
    Each SCP component accumulates its administrative and maintenance measurements for specified intervals (normally 15 to 30 minutes) and forwards these to the node manager on request. The frequency of these measurement messages is controlled by the node manager (currently the requirements range from five to 60 minutes). Each SCP component also keeps a limited history of these measurements so that if a switchover of node managers occurs, the new primary node manager may request each SCP component to resend the statistics required.
    The node manager combines the measurements received from all SCP components for sending to the SMS and for viewing by the craft terminal.

- Threshold Analysis:
  Threshold analysis allows the SCP to monitor the performance and status of components within the SCP. The first level of threshold analysis is performed at each SCP component. The node manager performs threshold analysis for those measurements requiring input from multiple components. When a threshold is exceeded, appropriate notifications and alarms are sent out.
  Threshold values are entered at the craft terminal or come from the SMS. Threshold values are checked at input and maintained by the node manager. New threshold values are distributed to the SCP components. An example of a threshold involves maintaining a count of the number of disk errors for an application. In this case, each host processor must be sent the value specifying the number of errors (threshold) after which the Node Manager is to be notified.
  Another example is the tracking of message response times that could trigger a call gapping event to be sent by the node manager to the overloaded application in the application host.
- Load Distribution:
  The node manager controls the distribution of incoming service messages from the SPIs to application hosts (Hosts) and of outgoing service messages from hosts to SPIs. The node manager maintains information about which hosts are available, the applications of each host, and which SPIs and signaling links are available. This information is distributed to the hosts and the SPIs initially and whenever changes occur. The SPIs use the host information to vary which host will receive an incoming service message, giving even distribution of the messages over available hosts. Hosts use SPI information to distribute outgoing messages over available SPIs.
- MTP/SCCP Management:
  The node manager contains MTP level 3 Signaling Network Management functions and SCCP management functions to support the CCS7 interface to the Common Channel Signaling Network.
  The MTP management controls the current message routing and configuration of the CCS7 Signaling Network facilities. In the event of change of status, the MTP also controls network reconfigurations and other actions to preserve or restore the normal message transfer capability.
  The SCCP management maintains network performance by rerouting or throttling traffic in the event of failure or congestion in the network.

Maintenance Functions: Maintenance functions include problem detection, notification and isolation, service recovery, verification, repair and error logging.
- Problem detection:
  The SCP hardware components have automatic error detection and correction logic features. Errors are logged for later analysis by field engineers, and are also reported as alerts to the node manager (via NetView) for analysis and recovery. Each SCP component reports anomalies in the operation of components other than the node manager, such as not receiving an expected response from another component. The hardware error detection logic handles

hard and soft storage errors and, in some cases, replaces an active processor
controller with a standby processor.

- Service recovery:
  The node manager redefines the SCP configuration to bypass a failing
  component (SPI, OSNI, or application host) and notifies other SCP components
  affected. Once a failed component is restored to service, the node manager is
  notified, and the SCP configuration is redefined to place the restored
  component online again.

- Problem notification:
  Each SCP component notifies the node manager when particular events occur.
  A message indicating trouble requiring maintenance personnel action causes the
  node manager to generate audible and visual indications to the local and/or
  remote alarm panels and an output message to the local and/or remote craft
  interface console. This alerts maintenance personnel to the problem, its type of
  severity, and the equipment affected.

- Alarm management:
  There are critical, major, and minor alarms. Critical alarms have the highest
  priority and are reported before any major or minor alarms. Major alarms are
  reported before minor alarms.
  Output messages resulting from problem notification follow the same ordering
  priority and include a time stamp and sequence number to indicate the order in
  which the problems actually occurred. If the number of output messages
  exceeds the capacity of the craft console, low priority messages are replaced by
  messages with a higher priority.
  The node manager maintains information about errors and their severity level.
  These can be changed using the parameter change control facility of the node
  manager.

- Verification:
  Each SCP component automatically verifies the existence of a problem by
  retrying a failing operation and/or running a test routine. It notifies the node
  manager of the result. When the node manager detects an anomaly, or when
  an SCP component reports the failure of another component, the node manager
  requests a test of the failing component or subcomponent to confirm that there
  is a problem.

- Isolation:
  This occurs once a problem has been detected and verified, and service recovery
  and protection have taken place. Problems are isolated online by the node
  manager to the SCP component level and by the SCP component to the
  subcomponent level. Online problem isolation permits logical removal of the
  failing part from normal operation by reconfiguring the system. Diagnostic
  programs and offline procedures by maintenance personnel also isolate
  problems.

- Repair:
  The SCP components are designed for continuous operation and can be
  serviced by technicians without the need for special tools. Many SCP
  components, such as the SPIs, OSNIs and craft terminals, can be easily replaced

by spare units. The first level of repair for these components is replacement and sending the failed unit to a central maintenance facility for diagnosis and repair. The returned unit then becomes a spare.

With mainframe processors, such as the application hosts and node manager, the failed SCP component is reconfigured out of the system, and diagnosis and repair are performed offline. The repaired unit can then be brought back online through the node manager.

- Error Logging:
  Error messages sent to the node manager are maintained for the number of days specified through the parameter change control facility of the node manager.

Configuration Control: Configuration Control consists of initialization, recovery, switchover, shutdown, and reconfiguration.

- Initialization:
  The first step in initialization is powering up all SCP components which causes each component to be Initial Program Loaded (IPL) and brought to an idle state. The node manager establishes communications with the OSNI and requests that it establish communication with the local and remote craft terminals to allow logon requests. The craft logs on to the primary node manager using appropriate procedures and password and sets up the operating parameters for the SCP. He then proceeds to bring up the other SCP components. The craft terminal (local or remote) used for logon becomes the controlling terminal.

  The primary node manager establishes communication with the other SCP components for node management activity. It distributes the operating parameters to each component, starts timer routines for status messages from the SCP components (for statistical summary records for SMS, and other purposes), and requests that the other components begin operation. It also establishes communications and coordination with the backup node manager. The SPIs establish communication with the application hosts for message interchange and align the active CCS7 links. They then process CCS7 messages incoming over these links.

  The OSNIs establish communications with the application hosts for database update inputs from the SMS, with the SMS (over the operation system network) for receiving database updates and SCP parameters, and for processing requests for SCP measurements from the SMS. The application hosts communicate with the SPIs and OSNI and begin processing messages and database updates.

- Recovery:
  Once a failed SCP component has been manually repaired and verified, the craft notifies the node manager that the component is now available. The node manager updates its configuration tables and notifies the other SCP components, the STPs and/or the SMS if necessary. The node manager then distributes the revised configuration parameters to the other SCP components so that the recovered component may participate in SCP processing.

- Node Manager Switch-Over:
  The node manager is configured with one processor designated as the primary, and one or more processors as backup. Other processors are potential node managers, to be activated when the primary or backup node manager is disabled or deactivated.
  The primary node manager acts as a focal point for all control, alarm, status and measurement activity. The backup has the same functions as the primary but is not the interface to the other SCP components. The backup is synchronized with the primary, to provide orderly switch-over.
  The backup node manager monitors the status of the primary through the periodic exchange of messages. The backup node manager becomes the primary if the primary node manager becomes disabled or at the direction of the craft.
- Shutdown:
  The craft may request the node manager to shutdown the SCP. The node manager notifies the other SCP components to prepare for shutdown; they complete processing, empty message buffers, and notify the node manager that they are ready for shutdown. When all components have shutdown, the node manager notifies the craft.
  When a planned shutdown of an operational system occurs, the operational system sends a message indicating that it is no longer active. Once this message is received, the node manager then starts to queue the messages. The activate message invokes the link initialization procedures and the queued messages are sent.
- Reconfiguration:
  The node manager assists the craft to configure or reconfigure the SCP. The node manager presents all SCP components to the craft and allows him to designate the active and non-active components. The SCP then validates the requested configuration, and notifies the craft of any discrepancies or anomalies. The process is repeated until a satisfactory configuration is specified. When the new configuration is accepted, the node manager starts the initialization process for the new components, and removes the components (craft designated). The process is performed systematically to minimize the impact on SCP performance.

Administration: These functions include security, measurement reporting, specification data management and generic program alteration.

- Security:
  Security consists of logon procedures, authorization, and enforcement processes. It requires the craft to enter a valid user ID and password before he can gain access to the node manager, and to functions restricted to specific users. Some files have their own password to prevent unauthorized access or modification.
- Measurement Reporting:
  The SCP generates reports daily, day-to-hour (on demand), hourly and every five minutes. Reports may be on demand, or scheduled. Reports information are maintained in the parameter control file.

A report is generated from the measurement data that has been collected, totalled and saved. It is sent to the location requesting the report or another requested location. The measurement reports may also be displayed online at the local or remote craft terminal.

- Specification Data Management:
  The node manager maintains node and application specification data obtained from the Signaling Engineering and Administration System (SEAS) and the Service Management System (SMS).
  The types of data maintained include link addresses, X.25 link address, public packet switched network addresses, application identifiers, destination point codes, and so on. Updates to this data may be made at the craft terminal or through messages from SEAS and SMS.
  The node manager maintains internal specification data consisting of elements that control SCP operation. These elements may be entered and changed using the craft terminal. This data contains user IDs and passwords, frequency of statistics collected per component, how long the node manager keeps statistics, how long the SPIs, OSNs and hosts must keep statistics, and so on. The node manager distributes specification data to the SCP components at initialization and when the components are changed.

- Generic Program Alteration:
  Software is maintained at an integration and test facility separate from the SCP. Software updates, verified for correctness, are distributed to the SCP either electronically or on magnetic tape. The node manager coordinates the distribution of the updates to the SCP components. The new modules are placed into a temporary library using standard maintenance transaction and software update procedures. The modules are put into production by recycling (re-IPLing in some cases) the host, OSNI, or SPI to the latest copy.

Operational Control: Operational control of the SCP consists of the:

- SCP Management Interface:
  The node manager is the user interface that allows the craft to manage the SCP and its components. The node manager receives and integrates information from the SCP components (node manager, SPIs, OSNIs, application hosts, and so on) for presentation at a single craft terminal. There may be more than one craft terminal. A local craft terminal at the SCP site is defined. An off-site remote craft terminal on the operations system network is also defined.

- SCP Maintenance Interface:
  The node manager includes a maintenance facility that allows the maintenance crafts, through local or remote terminals, to validate the operational capabilities of the SCP. This includes diagnostic testing of the hardware, testing links to the STP and operational systems, and testing out paths between the SCP components.

- Alarm Subsystem:
  The alarm subsystem contains lights and bells that are turned on when an alarm condition arises requiring manual intervention. A local alarm panel at the SCP site is connected to the OSNI and is accessed by the node manager. A remote

alarm panel at the central site has access to the SCP via the operations system network and is accessed by the node manager through the OSNI via the token ring. The alarms are turned off by the craft once the problem is corrected.

- Alert Output:
The node manager combines the output of the other SCP components with its own data to produce reports. The reports are displayed on the craft console, printed in hard copy, or sent to the appropriate destination.

- Diagnostics and Test Tools:
The hardware monitors errors in its SCP components. Any problems are sent to the node manager via NetView as hardware alerts. Maintenance personnel use offline diagnostics and error logs to diagnose and isolate failures to a more granular level and will replace the failed component(s).

- Continuous Automatic Tests:
Continuous automatic tests monitor active and standby equipment, confirm proper subsystem operation, and test the SCP's external message processing capability. Automatic problem determination tests are available to detect call processing, availability, and message processing irregularities.
Each SCP component periodically runs test routines to exercise and determine the proper operation of its subcomponents and interfaces. The results are reported to the node manager as part of the status reports. Errors may cause reconfiguration and alarms.

- Confirm Subsystem Operation:
The craft can verify that a particular SCP component is active by sending a request from the node manager to that component, asking it to perform an action that indicates that its systems are functioning. For example, to see whether a SPI is functioning, a request is sent to the SPI which responds to the message indicating that it is still active.
The craft may also request the SCP component to run diagnostics that detail its current configuration and whether it is operational. The node manager may also request these tests to verify a reported error.

- Test SCP external message processing capability:
The craft may verify that a particular SCP component can send messages to its STP or OSN counterpart by sending a request from the node manager to that component and requesting it to send a null transaction to a system over a specific link.

- Monitor Equipment Power Levels:
The primary power to the SCP hardware is continuously monitored for power loss. When a loss is detected, an alarm is raised.

4.4 Signaling Transfer Point (STP)

The STP may be a standard component of the underlying CCS7 network. General Bellcore STP requirements include STP functions for IN applications.

4.4.1 Functional STP Requirements

An STP in an CCS7 network must fulfill basic CCS7 requirements. Only requirements for IN applications, which need not necessarily be provided in a basic CCS7 network, are examined here.

SCCP Protocol Class 0 (and 1):   In an IN architecture, the STP must be able to act as a relay point for Signaling Connection Control Part (SCCP) protocol class 0 (the SCCP datagram service, without attempt for in-sequence delivery). To do this, the STP must implement or supply:

- Unit Data Message (UDT), to transfer information for the SCCP user
- Unit Data Service Message (UDTS), to report unsuccessful delivery of a UDT to the originating signaling point
- Global Title Translation (GTT), to map some kind of global address, such as an IN Service trigger, (for example, an 800 number) to the appropriate destination.

Protocol class 0 is also needed for SCCP management functions. Protocol class 1, which attempts in-sequence delivery, may be useful for some applications.

SCCP Management: SCCP management includes

- Destination Point Code (DPC) Status Management
  The status of point codes relevant to the SCCP must be maintained and GTT adjusted according to availability of replicated nodes.
- Subsystem Status Management
  The status of subsystems relevant to the SCCP must be maintained and GTT adjusted according to the availability of replicated subsystems. This is different from point code status management as a subsystem might be unavailable while the node on which it resides is still available.
- Traffic Mix Procedure (not a CCITT procedure)
  This procedure informs subsystems of what kind of traffic mix they are receiving.

Additional STP Functions: An STP serving as a traffic concentrator to SCPs may be added where the CCS7 network is not well suited to handling the additional IN traffic throughput and/or unable to provide the required reaction time.

If a vendor feature node (VFN) is incorporated in the IN architecture via CCS7, its access point (the STP) may need to provide special screening or security measures within the MTP and/or the SCCP. This feature may be necessary when the VFN is not within the jurisdiction of the CCS7 network service provider.

4.4.2 CCS7 Network Architecture and IN

STP Overlay Network: If an CCS7 network uses an overlay network of stand-alone STPs and therefore mostly quasi-associated signaling (that is, not paralleling the speech path), STPs should easily manage to accommodate additional IN traffic. Introduction of the IN still depends on available SCCP functions.

If all STP functions required for the IN are available as in the North American network, the IN can be introduced immediately.

If existing STPs do not implement all required functions, for example, GTT or management of replicated subsystems, an STP (stand-alone or integrated) with these functions should be introduced.

No STP Overlay Network: Where the CCS7 network does not use an STP overlay structure but uses associated signaling (that is, paralleling the speech path) the basic CCS7 network must be extended to accommodate IN traffic.

The Siemens STP for IN Applications

To introduce IN in CCS7 networks that do not have the SCCP requirements needed for IN applications, STPs with the functional and dynamic power can be used.

Siemens can provide an STP, stand-alone or integrated in an SSP or normal exchange. Where appropriate, EWSD switches already used in the network can be upgraded.

General STP architecture is shown in Figure 22. Figure 23 shows how the STP could be imbedded in the IN CCS7 architecture.

The Siemens stand-alone STP is based on the EWSD family of systems (which is also used as an SSP). It consists of an EWSD system and the Siemens CCS7 message switch, the CCNC.

This solution has the advantages of immediate availability, proven architecture, high reliability, and customer acceptance in many countries.

Where the expected traffic does not justify a stand-alone STP, a regular EWSD system already in the CCS7 network can perform the required functions.

The EWSD architecture is shown in Figure 24, and the components required for STP applications are described below.

Common Channel Network Controller (CCNC): The CCNC (see Figure 25) performs all MTP functions, and distributes traffic throughout the EWSD system.

The CCNC is a multiprocessor system consisting of:

* Multiplexing equipment (and modems) providing the physical interface to the outside world
* Signaling link terminals (SILT) providing the level 2 MTP functions
* A signaling link terminal controller (SILTC) providing an interface between SILTs and the CCNP
* The common channel signaling network processor (CCNP) providing the level 3 MTP functions.

The CCNP consists of:

* The signaling management processor (SIMP)
* Eight signaling periphery adapters (SIPA)
* The coordination processor interface (CPI).

Various components of the CCNC and the CP administer and maintain the CCNC.

Coordination Processor (CP): The CP coordinates call processing functions, administers EWSD, and provides peripheral interfaces. The coordination processor is used for STP operation, administration and management and to interface with the operation, administration, and management periphery. Figure 26 shows the CP113, the most powerful coordination processor in the EWSD family.

Switching Network: The switching network performs the physical switching functions in EWSD. In the STP, it connects the signaling channels entering the system via the line trunk groups (LTGs) to the CCNC.

Line Trunk Groups: The line trunk groups perform some call processing functions, controlled by the CP, and interface EWSD to the transmission links.

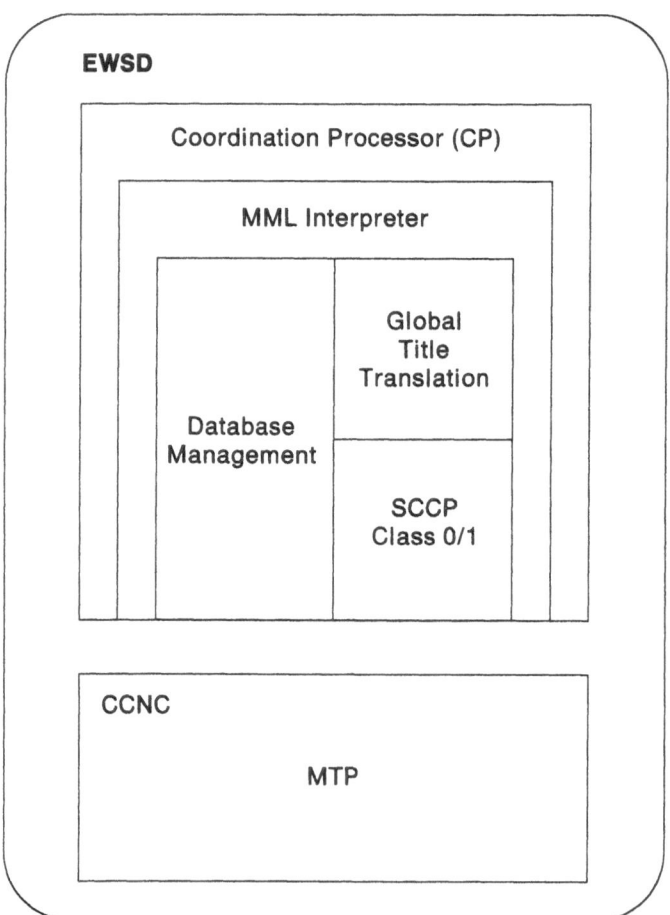

Figure 22. General STP Architecture for the EWSD

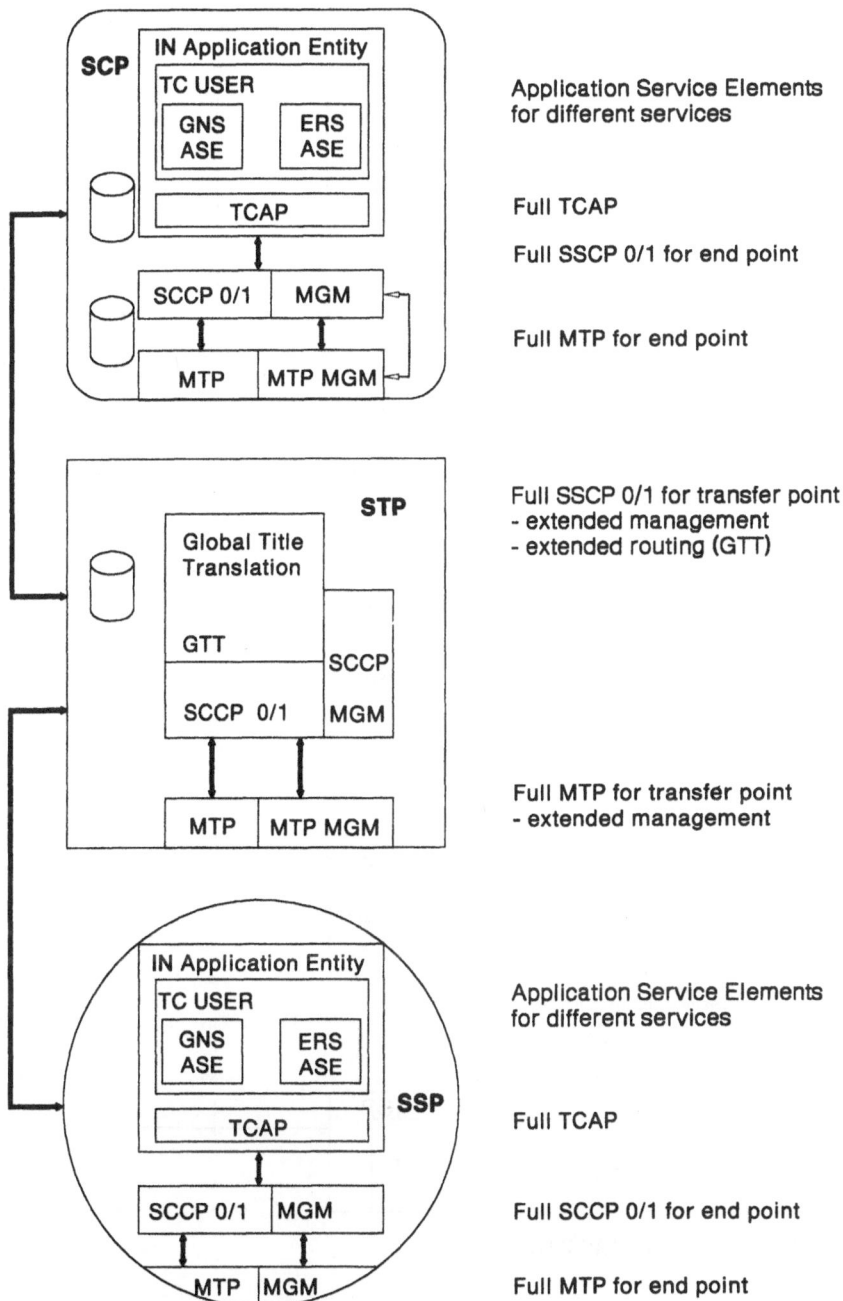

Figure 23. Possible IN CCS7 Architecture for the EWSD

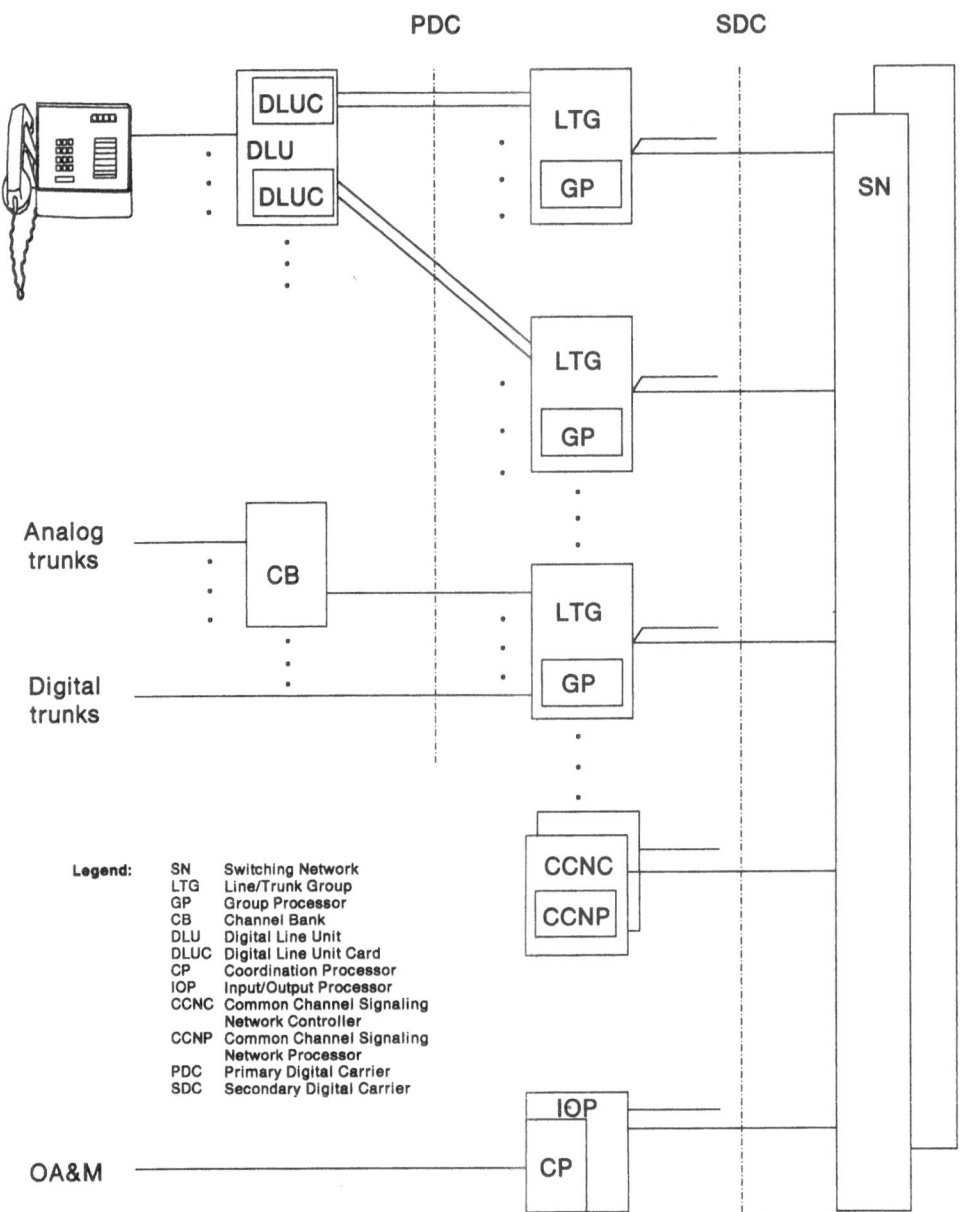

Figure 24. Architecture for the EWSD

Software

Software is distributed between the coordination processor (CP) and the LTG processor (GP) to achieve a high degree of parallel processing. The structure within each processor is the same to achieve a high degree of parallel processing. A particular subsystem group, the operating system, is shown in Figure 27.

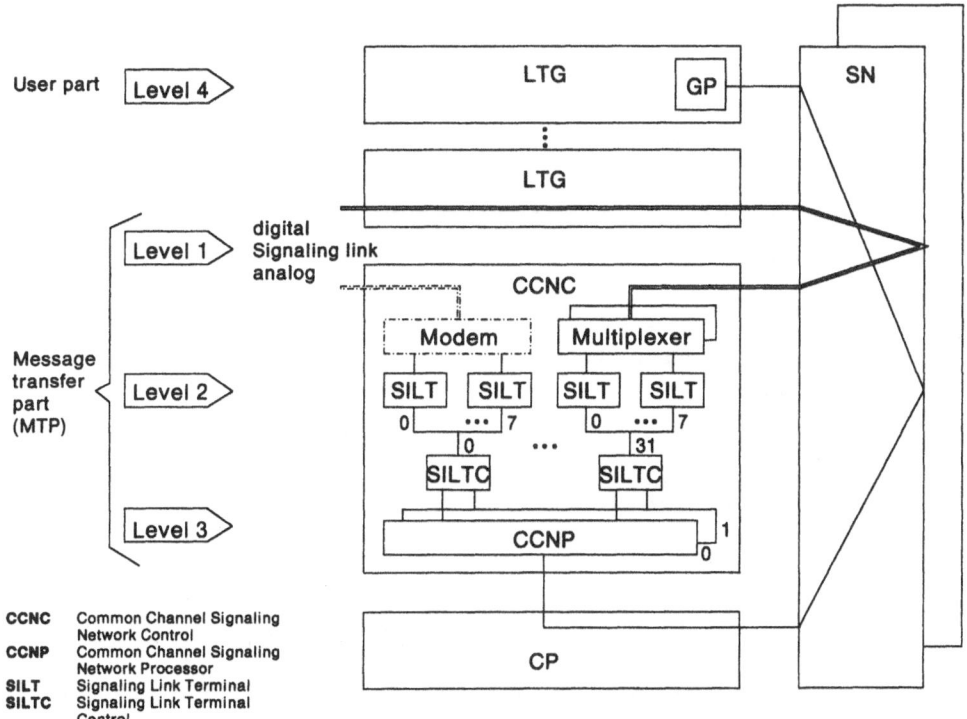

Figure 25. Common Channel Network Control for the EWSD

CCS7 Interface

EWSD supports both CCITT CCS7 and ANSI CCS7 formats and procedures, such as:

- Normal and emergency alignment
- CCITT and ANSI formats for signaling point codes and signaling link selection fields
- CCITT and ANSI timers for levels 2 and 3 (where required).

EWSD CCS7 implementation is continuously updated to comply with national and international recommendations. Siemens participates in national and international standard-setting organizations for CCS7 development. Such participation guarantees that the EWSD CCS7 implementation fulfills market requirements.

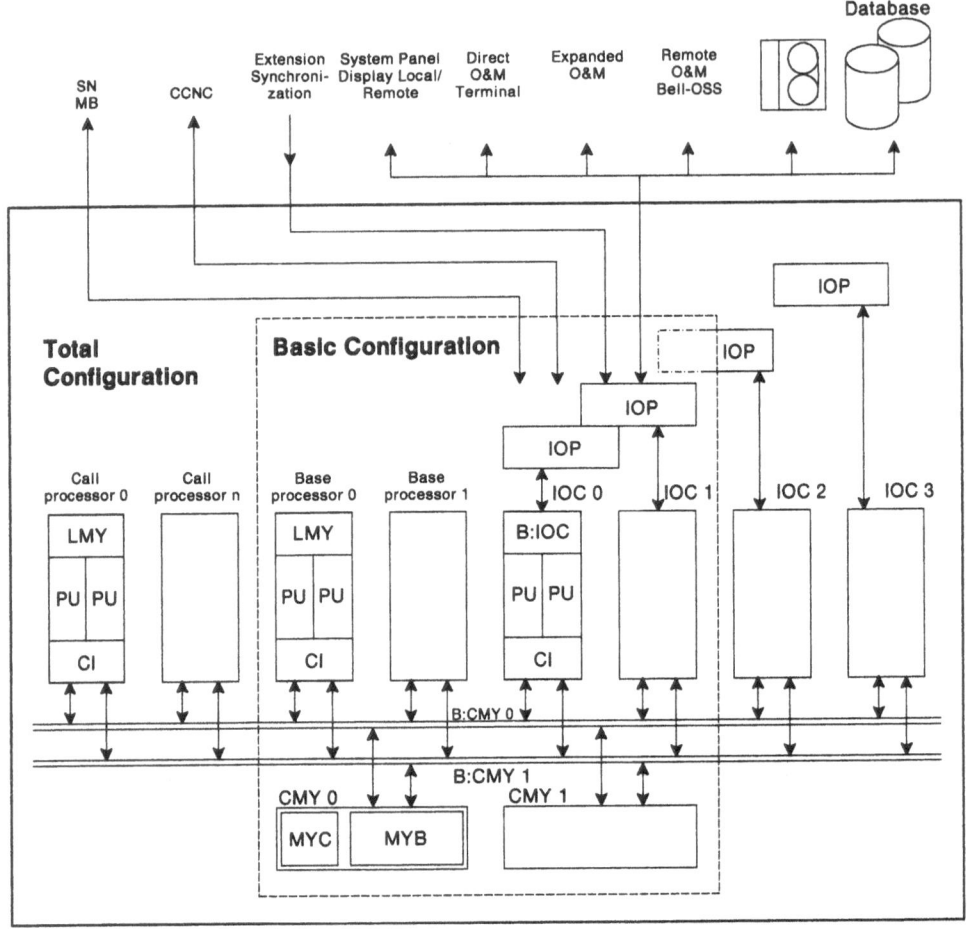

Figure 26. CP113 Coordination Processor for the EWSD

4.5 Service Switching Point (SSP)

With the SSP function, an exchange can interface with a database at a Service Control Point (SCP) for the performance of IN services.

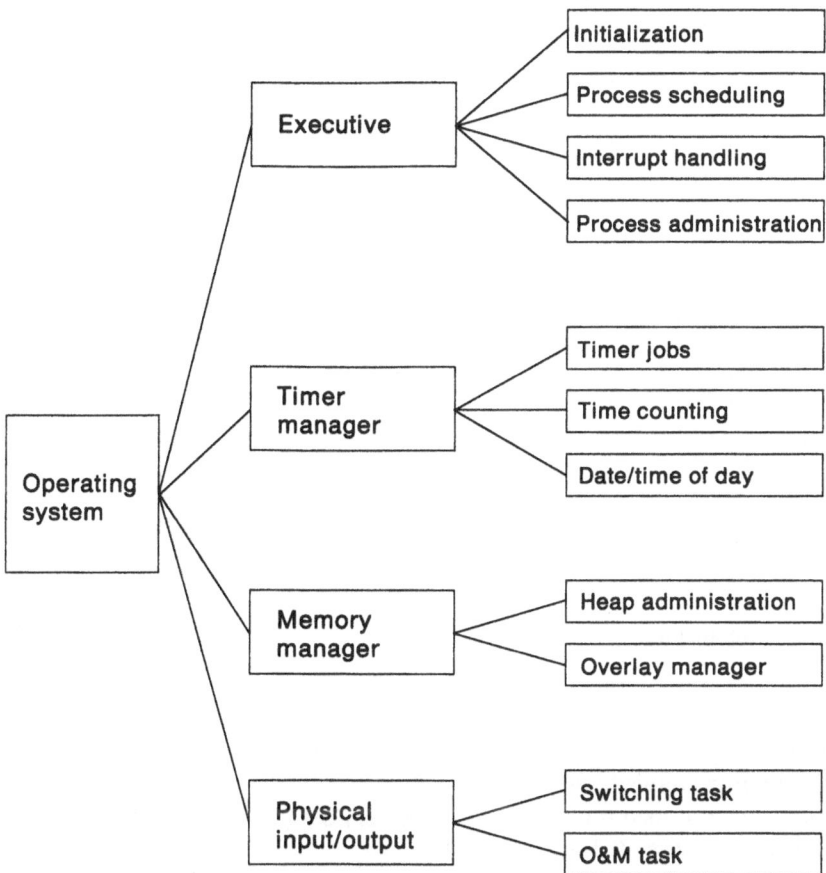

Figure 27. The Operating System

The SSP also provides normal switching functions. The SSP is comprised of the following components (as shown in Figure 28):
- Application parts which define the services that are supported
- A Transaction Capability Application Part (TCAP), which provides a part of the application layer protocol.
- A network layer, which defines the communication network for which the protocol is designed. The SCCP is currently defined as the network layer, based on the CCS7 MTP.

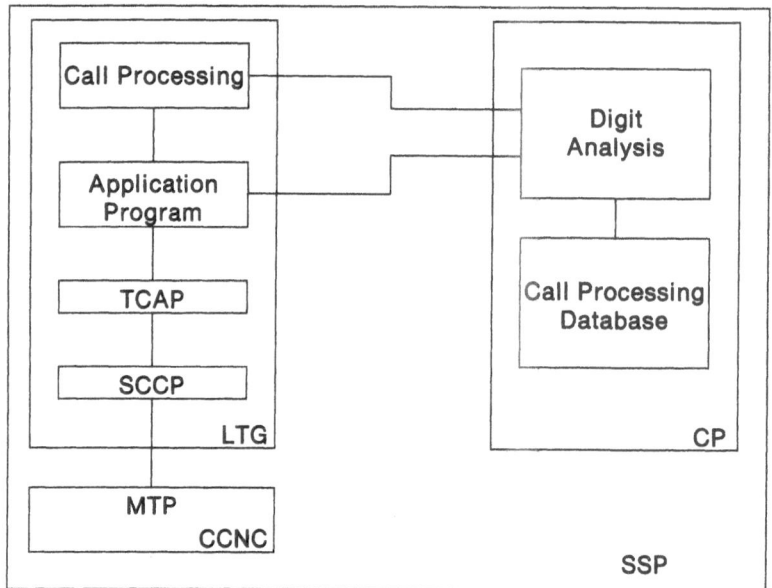

Figure 28. SSP Components

Signaling Connection Control Part (SCCP)
The SCCP provides a transfer capability for Signaling Data Units with or without using logical signaling connections.

Services are thus grouped into connectionless Services (Classes 0 and 1) and connection-oriented Services (Classes 2 and 3).

SCCP users are identified by subsystem number (SSN). The SSN field is one octet long.

Network specific subsystem numbers are assigned in descending order, starting with "11111110" (FE Hex).

For the initial application, only connectionless services (Classes 0 and 1) are required.

SCCP Management: maintains network performance by rerouting or throttling traffic during network failure or congestion.
Point code status management
Updates routing and translation information due to network failure, recovery or congestion indications received in the MTP-PAUSE, MTP-RESUME, or MTP-STATUS primitives.
Subsystem status management
Updates routing and translation tables so that the network can adjust to, and compensate for, failure, withdrawal, congestion and recovery of a subsystem.

Transaction Capability Application Part (TCAP)
The TCAP provides procedures to support a variety of applications, thus avoiding the inefficiency of specifically tailored procedures. Hence, the TCAP provides a framework for a common approach to new services within a network and for the service architecture for inter-network services. The TCAP is divided into two sublayers: the transaction portion, and the component portion.

Formatting of the TCAP protocol elements is derived from the Abstract Syntax Notation 1. The TCAP uses standard formatting rules, therefore the common formatting procedure (identifier, length and value) serves all TCAP users.

Since some parameters are common to more than one user, a common routine to build these parameters reduces the development and testing effort.

With the transaction sublayer, users can exchange information in a structured or unstructured way. Currently, the only user of the transaction sublayer (the component sublayer) enables its users (the applications) to invoke operations and report their results by exchanging components. It also allows the applications to structure or control the information exchange via services provided by the transaction sublayer.

Four components that are currently defined are:

- **Invoke**
  Used to invoke an operation. Invoke only contains one operation, but can also carry the response to a previously received Invoke. "Not Last/Last" indicates whether or not further responding components are expected.
- **Return Result** (RR) (Not Last/Last)
  Reports a success and returns the result of an invoked operation.
- **Return Error** (RE)
  Reports that an invoked operation could not complete successfully.
- **Reject**
  Reports the receipt of a message causing a protocol error at the transaction or component level.

Application Part
The application program interfaces with the switch functions and TCAP. The application program is triggered by switch functions or TCAP messages received from the remote end. The application program transforms instructions from the SCP into switch functions. Actual instructions and functions are service-dependent.

SSP—SCP Interface

TCAP Message from SCP: The requirement in this area is to map TCAP operations and parameters into existing routines and arguments of the switch, so that the function can be accomplished. The results of the switch's routines map into Return Results or Return Errors.

TCAP Message to SCP: The requirement in this area is to map switch functions and arguments into TCAP operations and parameters so that the TCAP component can be built and sent to the SCP.

The Siemens Service Switching Point

Like the STP, the Siemens SSP is based on the EWSD family of systems. The EWSD-SSP hardware is functionally identical to the EWSD hardware.

IN Specific Software: No special software platform is provided for IN/1 applications. For each IN service, a special service module is created which coordinates the interaction between the SCP and the switching function.

The basic software architecture is demonstrated using the GNS service as an example. Figure 29 shows the system structure and the information/message flow. When a subscriber dials 800-NXX-XXXX, the following process takes place:

1.   The switch-function program (call processing) in the LTG collects the digits and sends the digit block to the digit analysis program in the CP.
2.   The digit analysis program detects that the call is an 800 call, collects the destination related information from the database and passes it to the service application program (800-service) in the LTG. This is the trigger for the application program.
3.   The 800-service program verifies the DPC status. It assigns the transaction ID and the invoke ID for the provide instruction (operation), formulates the TC component invoke and passes it to TCAP.
4.   TCAP formulates the query, provides component and transaction level processing, and passes to SCCP. TCAP also provides the association at transaction level and the correlation at the component level.
5.   The SCCP forms the SCCP header. The SCCP passes the message to the MTP.
6.   The MTP sends the message based on the DPC to the remote end (that is, to the STP for GTT, or directly to the SCP).
7.   The STP performs the global title translation, and the new value of the DPC (that of the SCP) is placed in the routing label. The SCCP called party field is modified to reflect that routing is now based on the SSN, and the value of the SSN is inserted or replaced by the STP. It then sends the message to the SCP.
8.   The SCP processes the query, and the result of the translated number is placed in the "Connect Call" invoke. The SCP may also request the "Terminating Information" which is placed in a second Invoke. The SCP sends these components in a response package directly to the SSP. (The SSP point code is derived from the SCCP calling party field).
9.   When the SSP receives this message, it passes through MTP, SCCP and TCAP before arriving at the application. The MTP and SCCP provide appropriate processing at their levels. The TCAP terminates the transaction and delivers the message to the application.
10. The application program provides the translated number to the switch function, and normal call processing continues.
     Note: If the SCP requests terminating information, the 800-service application indicates to the switch function that the application should be notified when the call is over with the duration information. The application stores the SCP's DPC as well as the ECHO DATA field received in the message for later use.
11. When the subscriber has completed the call, the switch function checks whether the application should be informed. In the case of terminating information, it sends an event to the application.

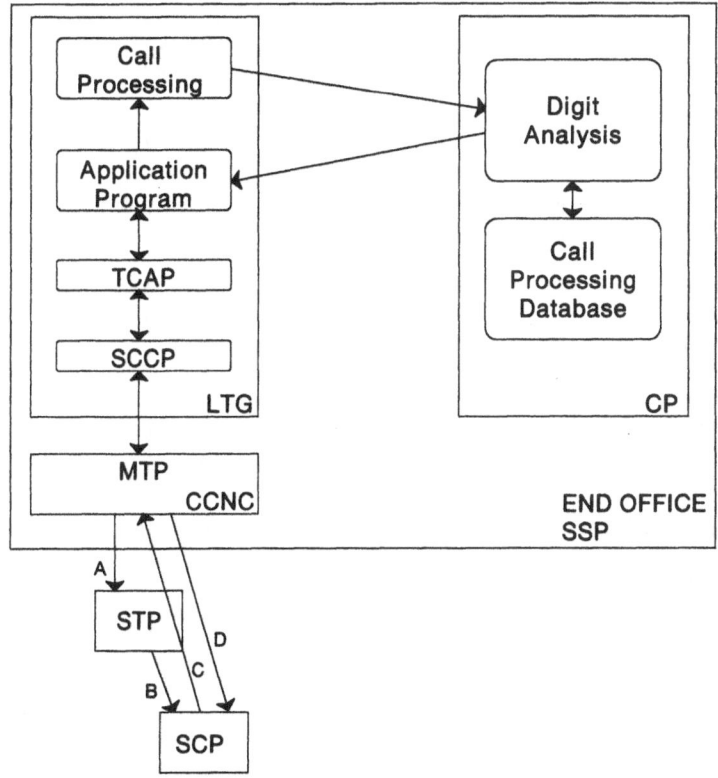

A	Query with Invoke. Provide instruction (DPC - STP, Route according to GTT)
B	Query with Invoke. Provide instruction (DPC - SCP, Route according to SSN)
C	Response with Invoke. Connect call, and Invoke. Send terminating info (DPC - SSP, Route according to SSN)
D	Unidirectional package containing terminating info (DPC - SCP, Route according to SSN)

Figure 29. EWSD System Structure and Information and Message Flow for the GNS

12. The application formulates "Return Result" with terminating information and places it in an "UNIDIRECTIONAL" message to be sent to the SCP directly (without GTT request) using the original transaction ID.
13. This message passes through TCAP, SCCP and MTP (CCNC).
14. This message is received by the SCP and the function is complete.

Operator Interface: Commands from the craft person provide the trigger indication and destination address to which the request for service-related instructions is to be sent in the database. Three commands are required; "create", "cancel", and "display a point code group". The following parameters are required:

- Trigger indication—Identifies the service to be invoked.
- CCS7 Destination address—DPC of the remote end to which the query can be sent.
- GTT Indicator—A one bit field indicating whether or not a GTT request is to be generated.

• Primary/Backup—Identifies the primary or back-up SCP/STP pair.
The craft person must provide this information in the "create" command. The DPC must be verified against the CCNC database, as the routing must exist in the MTP for the entered DPC.

The system maintains the DPC status. For the display command, along with the above information, the status of the DPC is also displayed.

Inaccessibility to primary and backup DPCs is monitored by the system and reported to the administration system if their status changes.

Administration: Measurement counts are kept at TCAP level. Application related counts are defined based on the service being offered. They are collected periodically and reported to the collection centers.

Flexibility of the EWSD SSP Design: The SCCP (Classes 0 and 1) and TCAP are implemented in EWSD. The SCCP is implemented so that it can serve existing users and allow for the addition of new users. The SCCP supports sending messages direct to the SCP and requesting global title translation (GTT) of the STP. The TCAP design uses a general formatting concept to build a TCAP message, so that new applications can be added with minimal changes to TCAP.

4.6 Future Considerations

IN/2 Functional Elements
The current architectural work on IN/2 depends on new and improved logical modules and network elements. The logical modules are:
• SSP/2
• Network Information Database (NID)
• Network Resource Manager (NRM)
• Adjunct Service Point (ASP)
• Service Logic Interpreter (SLI).
The modules reside at the following network elements:
• Service Switching Point (SSP)
• Service Control Point (SCP)
• Intelligent Peripheral (IP).
Communication between the logical modules takes place via service independent requests and responses carried in messages via open interfaces.

Functional Components in IN/2

A major advancement towards service independent communication between the SSP and SCP is the creation of new functional components (FCs).

FCs are modular operations; combinations of FCs provide complete network functions, called Service Logic Programs (SLPs). SLPs can be used to create and customize IN services. An SLP is composed of a set of FCs, and IN services are comprised of a set of SLPs.

The introduction of the Service Logic Interpreter (SLI) will allow the processing of the service logic programs.

Part 2. Green Number Service

This part provides detailed information about the Green Number Service. It examines what the service is, how it operates in the IN, and the factors to consider when implementing GNS.

This part is GNS-specific; it adds to, and does not replace "Part 1. Overview of the Intelligent Network" of this report.

Chapter 5. GNS Service Description

5.1 Overview

Basic Service
Calls containing a special access code (for example in the USA 800, in West Germany 130) trigger the interrogation (via CCS7) of a database to determine the real (target) number. The real number is used for call setup.

Enhanced Service
Enhancement of the basic service may depend on factors like:
- Origin (Automatic number identification (ANI), area code, region)
- Time and date
- Office status; open or closed
- Caller—network dialog to exchange additional information for determination of the real number.

Background
The GNS originated in the USA as the "800 Service". It was first deployed in 1967 by the Bell System, using special toll screening exchanges. In 1981 CCIS was used to support the 800 service using a centralized database concept (Network Control Point). The GNS has enjoyed tremendous success; in 1986 it generated $5 billion revenue, with a 20% growth rate in the number of calls annually.

Benefits
The benefits of GNS are:
- Service user (person who dials the GNS Number)
  - Calls are free
  - Speed and ease of use
  - No waiting for operator assisted procedure
  - Same number everywhere is easy to remember
  - Extended hours of operation (avoid busy hour, busy tone)
  - Wide availability.
- Service subscriber (owner of the GNS number)
  - Stimulated business by increasing sales and reaching new customers inexpensively

- Extend market coverage for products and services
- Possibility of numbers with advertising relevance
- Reduce sales, services, and communication expenses
- Registration of ANI which is communicated to the subscriber
- Flexibility to direct calls to best answering location
- Improve internal communications by monitoring the sources of, and reasons for calls.
- Network operator
  - Increase network use, and thus revenue
  - Increase network efficiency, and thus profit
  - Valuable customer service improves image
  - Faster response to network demands from customers
    - Number introduction
    - Feature modification.

5.2 Functional Description

General

When a service user (the caller who dials the special access code) makes a GNS call, he is connected to the closest SSP (an end exchange or a tandem exchange). In the SSP, a query is triggered to the database located in the SCP. The query is transmitted to the SCP via the CCS7 network, including STP nodes.

The GNS application program in the SCP reads a record from the GNS database. The key for the record search is the service subscriber's number, which was included in the query (the service subscriber is the company or person that subscribes to the GNS). The GNS application program translates the dialed (virtual) number to a real number, according to service parameters previously established by the service subscriber. The target number combined with service logic instructions is sent to the SSP via the CCS7 network and STP nodes. The SSP and SCP communicate using the TCAP-protocol.

Having received instructions and the target number, the SSP completes the call setup, routing the call through the network to the appropriate destination (for example, the sales exchange of a company).

The enhanced GNS determines the final destination number in a dialog with the service user, using an announcement machine and a touch tone instrument or an ISDN terminal. (For example, "Dial 1 if you want reservations", "Dial 2 if you want information".)

The service subscriber can specify a customized GNS using the SMS via a dial-up connection and a data terminal. Transaction processing between the SMS and SCP transfers the subscriber-specified data to the SCP, where it is used in GNS calls.

Service User's Perspective
GNS provides the service user with some service specific functions:
- Uniform access codes (per country, eventually European or worldwide).
- In many cases, the access code corresponds to the number describing the service (for example, the German Service 130 has the access code 0130).
- A GNS subscriber (for example, a large company) normally uses only one number per application. This number is usually short and easy to remember, or there might be a special GNS directory for the convenience of the service user.
- GNS is either free or the cost of a local call.
- To fulfill specific requests a dialog between the service user and the network may take place using the voice announcement and input facilities of the terminating equipment.

Service Subscriber's Perspective
The network operator assigns the GNS subscriber number considering special customer requests. The number remains unchanged if he moves, or if his real telephone number is changed.

The service subscriber must specify to which real number a service user should be connected, depending on user origin location, time of day, day of the week, and so on. In practice, this specification can be done in one of the following ways:
- The service subscriber is connected via a leased line or a dial-up connection to an SMS, and uses his terminal to change the specifications as desired.
- The subscriber submits a written service request to the network operator, which then updates the subscriber's record, via a service administration center.
- The subscriber calls the network operator service center and orally notifies an operator about the changes desired. The network operator personnel update the subscriber's record via a service administration center.

The service subscriber can choose the date on which changes should become effective. Using traffic statistics data, the service subscriber can determine the best routing lines depending on user needs and his own organizational and administrative resources, such as:
- Weekend stand-by services
- Connecting the user to the branch exchange closest to his home
- Not accepting calls from specific locations
- Unlisted GNS numbers can be used by company employees that need to call the company instead of calling cards. (In the US a large percentage of GNS numbers are unlisted numbers.)

Relation to ISDN Supplementary Services
The service "Reverse charging" may to some extent substitute or support Basic GNS. On the basis of an ISDN network, GNS may be extended to non-voice calls.

5.3 Standards

Service
The GNS is described using available Bellcore information and information known about other GNS implementations (for example, Service 130 in West Germany).

CCS7
Global Title Translation (GTT) may be required in the STPs if the GNS database resides in more than one SCP (see Figure 7).

Trunk Signaling
Normal interexchange signaling is required for GNS and for enhanced GNS. The trunk signaling should be able to forward to the SSP the origination of the calling service user.

5.4 Service Interaction

Service User
The service user dials an access code and a virtual subscriber number, using a POTS telephone or ISDN-based instruments. He is then connected to the target number. With an enhanced GNS, the service user may be prompted via an announcement machine to give additional digits or to select a number from a list of alternatives. The service user is guided via announcements. (If the call is made from a pay-phone the money may be returned.)

Enhanced GNS can only be used by DTMF or ISDN users, because dial pulsing cannot be used to convey information once the call has been set up.

Service Subscriber
The SMS provides a user interface with a hierarchy of menus to support and guide the service subscriber.

When interacting with the SMS, the service subscriber may be required to enter such data as:
- Security passwords
- Alternate billing number
- IC prefix
- Routing parameters
- Screening parameters (for example, calls from certain areas are restricted)
- Real number.

Access Instrument
No special equipment is needed to subscribe to GNS, however, if the service subscriber wants to update his own service parameters directly, he must have a data terminal or a PC that allows him to establish a leased line or a dial-up connection to the SMS. Access to the SMS is supported via both asynchronous and synchronous terminals and with either dedicated or dial-up facilities.

5.5 Billing

SSP
The originating exchange (SSP) must be able to generate AMA records for all TCAP GNS calls using the billing information received from the SCP database. The billing indicator field is a mandatory field in the response and therefore an AMA record should be generated for all answered calls. It should be possible to record unanswered calls.

The billing data collected at the originating SSP is formatted into records, which are automatically transferred to a service center that generates the actual customer bills.

Since the IN-based GNS can coexist with the traditional GNS, it is helpful if traditional GNS calls can be billed in the terminating exchange on a per service subscriber basis.

SCP
The SCP must provide billing functions for GNS calls. The network operator may choose to obtain GNS statistics from the SSPs with respect to GNS usage, and whether billing records are generated correctly in the SSPs. This collection process implies the ability of the SSPs and the SCP(s) to exchange GNS billing information. The SSP must be able to record and deliver to the SCP billing data such as call-event times and counter values. The SCP must record and supply the SSP with information on service features. (This will be via GNS specific TCAP messages described in the next chapter.)

SMS
The SMS must generate information about the service subscriber's profile (such as record size, number of changes, priority changes) to the network operator in order to generate billing information.

As well as paying for being able to update his record, the service subscriber may pay a monthly service subscription fee.

5.6 Service Logic

5.6.1 Distribution

SSP

The SSP database should contain originating triggers that enable it to detect calls that need a special routing service.

The SSP must contain logic to detect this trigger upon translating the incoming number. Logic is also needed to send an inquiry message to the SCP with the virtual number dialed and service user origin, and when the SCP has replied, to complete the call set up. Service logic must also exist to handle other types of messages from the SCP, such as Automatic Call Gapping (ACG).

The IN service logic must be able to coexist with traditional GNS logic.

SCP

The SCP should contain service logic to translate an incoming GNS number to a real destination number, according to the individual service subscriber parameters, and to handle exceptional situations (such as overload situations, database error situations, and "no number assigned").

Triggers

The trigger table in the SSP is administered by the network operator (not via SMS). It describes which part of the call data will trigger a query to the SCP, and at which point in the call the query is started. The triggering data in GNS is the number called (for example the 800-number 800-NXX-4567). The query starts as soon as the complete call address is received.

5.6.2 Functional Flow

SSP - SCP

SCCP protocol class 0 (connectionless, no message sequencing required) is used for all GNS messages. The following TCAP messages (components) flow between the SSP and SCP[23]:

SSP to SCP

.INVOKE	Initial Query from SSP (Provide Instructions)
.INVOKE	Procedural, Report Error (the SSP has detected an error in the SCP's response to the original "Provide Instructions" message)
.REJECT	Protocol error detected in the TCAP message (for example, unrecognized command)
.RETURN RESULT	Termination information (for example, billing information as response to a Send Notification message from the SCP).

---

[23] Service Switching Points, Bellcore TR-TSY-000064, FSD 31-01-0000.

SCP to SSP

.INVOKE	Connection Control (contains routing information)
.INVOKE	Network Management (Automatic Call Gapping)
.INVOKE	Caller Interaction (Play Announcement, for example an out-of-band condition)
.INVOKE	Send Notification (request to the SSP to send termination information)
.RETURN ERROR	An error was detected in the original Provide Instruction message data so that the requested operation could not take place (for example, unexpected data values)
.REJECT	Protocol error detected in the TCAP message.

For a detailed description of Bellcore defined GNS messages, parameters, and data elements, refer to "Service Switching Points", Bellcore TR-TSY-000064, FSD 31-01-0000. Figure 30 depicts the following GNS functional flow, which consists of four steps:

1. The service user dials the GNS access code which triggers in the SSP a query to the SCP. The TCAP message sent from the SSP to the SCP (INVOKE) contains the caller's number as well as the number called.
2. The SCP accesses the GNS database to get the service subscriber record corresponding to the number called (the key used to access the database is the GNS number called). The GNS application program determines the "real" telephone number of the service subscriber using subscriber specific parameters.
3. The application program requests the SCP communication program to create an INVOKE message which contains the "real" subscriber number.
4. The SSP completes the call based on this real number.
   Table 2 shows a detailed list of TCAP messages.

SCP - SMS

The generic messages sent between the SCP and the SMS are described in SCP - SMS General Interface Specification, Bellcore TA-TSY-000365. For a detailed list of these messages see Table 3.

5.7 Traffic Measurement Requirements

SSP

The following measurements are general to **Number Services**[24].
EADAS/NM Data:

• Number services SCP-initiated control discrete.
• Total originating number services attempts.

---

[24] Service Switching Points, Bellcore TR-TSY-000064, FSD 31-01-0000.

- Number services calls blocked for excessive calling to vacant codes or from nonpurchased NPAs.
- Number services calls blocked by SCP overload controls.
- Number services calls blocked for mass calling controls.

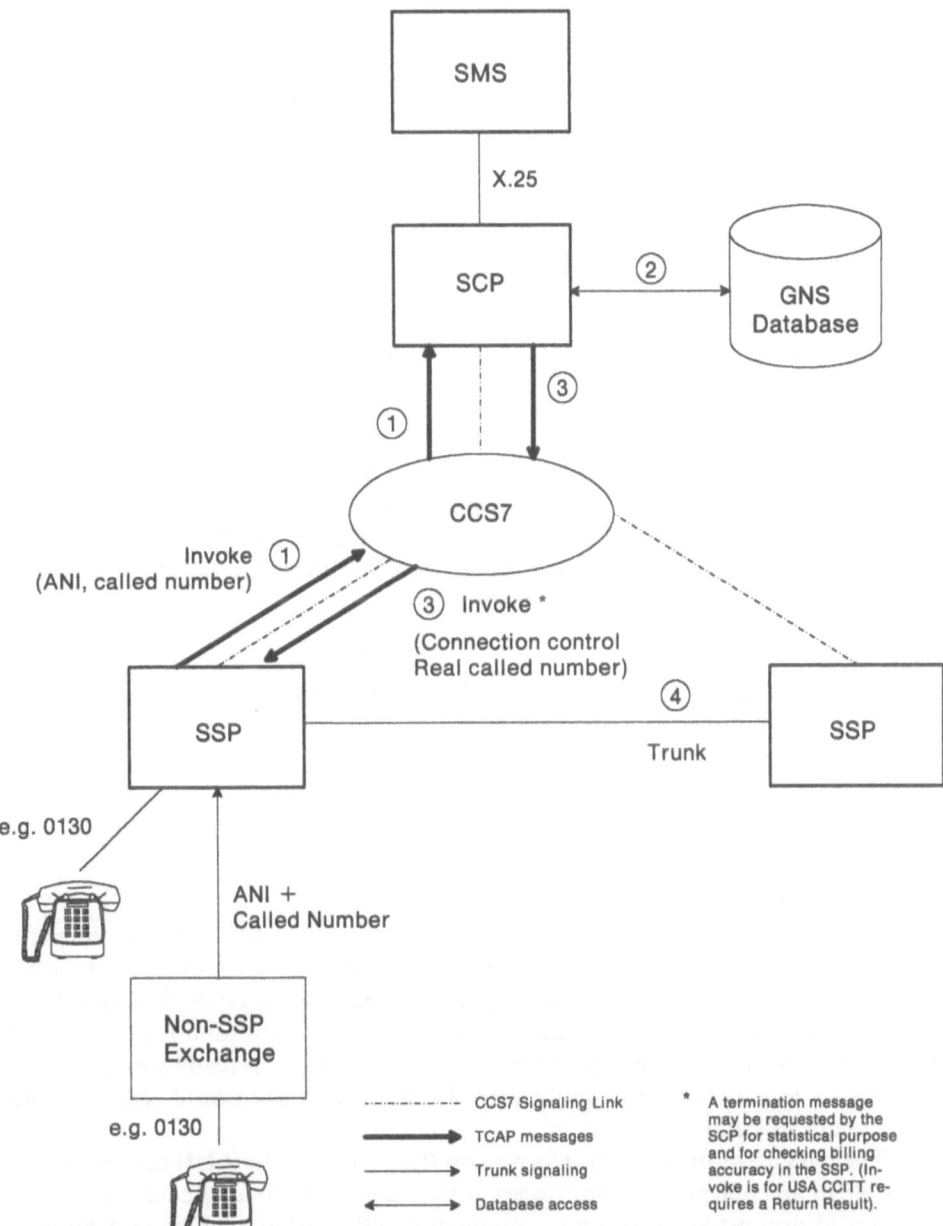

Figure 30. GNS Functional Flow—GNS without Prompting Option

- Number services calls blocked by SMS-initiated controls.
- Counts for control lists overflow.
  - Six-digit vacant code.
  - Ten-digit vacant code.
  - Nonpurchased NPA.
  - SCP overload.
  - Mass calling.
  - SMS initiated.

SCP

The following measurements are made:
- Number of GNS calls
- Number of stations busy
- Number of calls not answered.

5.8 Dynamic Requirements and Performance

SSP

The GNS user program should begin timing for a response from the SCP after sending a query message. A nominal value of three seconds should be used. If a response from the SCP is not received in this period, a reorder tone or an announcement should be given to the calling service user and the TCAP resources assigned when the query was sent should be released.

SCP

On receipt of a query from the SSP, the SCP should respond within 0.5 seconds on average, and within one second 99% of the time.

5.9 National Dependencies

Psychological factors may have to be overcome before the GNS can be successfully introduced in certain countries. For example, in countries with no detailed message recording (AMA), the service user may be concerned that a GNS call is not free of charge. There is no way of checking this until he receives his monthly bill and even then it may be difficult or impossible to determine if the call really was free of charge. An announcement could be provided that the call is free of charge before the actual call setup.

The passing of origination information (for example, the ANI) may not be possible in all countries. In these cases, origin dependent routing cannot be performed or must be based on the originating SSP and/or trunking origination information provided in the TCAP message.

The GNS may vary from country to country. For example, a PTT has identified the following GNS requirements:

- Routing according to:
  - Time of day, day of week
  - Origin (different granularity)
  - Traffic dependent overflow to alternate numbers, call distribution, and overload strategy.
- Overflow to network operator facilities.
- Personal identification number to be supplied before a call can be completed.
- Display on the service subscriber's equipment of the GNS number that was called.
- Display of billing information online for each call on the service subscriber's equipment.

5.10 Future Considerations

The billing function in the USA is done in the originating exchange. In the future it may be desirable to perform the GNS billing functions in the SCP. This will add considerably to the number of TCAP messages sent from the SSP to the SCP (one extra message per call).

It is expected that interactive GNS capability, where the service user is prompted by announcements to select a certain number, will be required by many network operators in the future.

Chapter 6. GNS Application Description

The description in this chapter is based on IBM's and Siemens' views on:
*   Functional requirements and allocation. Requirements are specified and each item is assigned to the SCP or the SSP depending on where it must be developed.
*   Functional units. Processing functions needed to handle the requested tasks are described and grouped logically into functional units for the separate components.
*   Administrative Units. Processing functions not directly required in the real time call handling process are combined into administrative units, separately for the SSP and SCP.

A bold printed SSP means that the associated function must be performed in the Service Switching Point.

A bold printed SCP means that the associated function must be performed in the Service Control Point.

6.1 Functional Requirements and Allocation

Real Time Call Handling
1. In the SSP a query to the database which is located in the SCP, will be triggered. For this purpose the SSP needs a database which contains (originating) triggers enabling it to detect that a call needs a special routing service.   **SSP**
2. The query is extended with TCAP protocol information and transmitted to the SCP via the CCS7 network and STP nodes.   **SSP**
3. The GNS application program in the SCP will read a record from the GNS database. The key for the record search will be the user-dialed service subscriber number which was included in the query.   **SCP**
4. The GNS application program will translate the dialed (virtual) number to a real number according to previously established service parameters like time of day, day of week, and so on.   **SCP**

5. The target number combined with service logic instructions and enhanced with TCAP protocol information is sent to the SSP via the CCS7 network and STP nodes.   **SCP**
6. Having received instructions and the target number, the SSP completes the call setup, routing the call through the network to the appropriate destination. **SSP**
7. If the SCP has requested "Call Termination Information", the SSP must send it as soon as the GNS call is terminated.   **SSP**
8. The GNS application program receives and processes the TCAP message containing the "Call Termination Information".   **SCP**
9. The GNS user program should begin timing for a response from the SCP after sending a query message. A nominal value of three seconds should be used. In case of timeout, a reorder tone or an announcement should be given to the calling service user and the TCAP resources assigned when the query was sent should be released.   **SSP**
10. These are the TCAP messages from the **SSP to SCP:**

    INVOKE   Initial Query from SSP (Provide Instruction)
    INVOKE   Report Error
    REJECT   Protocol error detected
    RETURN RESULT   Termination information.

11. These are the TCAP messages from the **SCP to SSP:**

    RETURN RESULT   Response containing routing information
                     (In the US this is done with Invoke (Connection Control))
    RETURN ERROR   Error in data detected (for example, no number assigned)
    REJECT   Protocol error detected
    INVOKE   Automatic Call Gapping
    INVOKE   Play announcement.

SMS-SCP Transactions

1. Transaction processing between the SMS and SCP transfers the subscriber specified data to the SCPs.   **SCP**
2. The transferred service subscriber specific data ("routing tree") is used to update the subscriber's record in the SCPs.   **SCP**
3. The SMS requests data from the SCP to generate reports.   **SCP**

Traffic Measurements
Note: This section is based on information from Service Switching Points, Bellcore TR-TSY-000064, FSD-31-01-0000.

*   Traffic statistics (for example, number of calls, number of busy situations, number of calls not answered) must be collected to be supplied to the network operator and service subscriber.   **SCP**
*   The following traffic measurements are general to number services, and must be provided in the GNS application: **SCP**
    - Number services SCP-initiated control discrete
    - Total originating number services attempts
    - Number services calls blocked for excessive calling to vacant codes or from nonpurchased NPAs

- – Number services calls blocked by SCP overload controls
- – Number services calls blocked for mass calling controls
- – Number services calls blocked by SMS-initiated controls
- – Counts for control lists overflow
  - – 6-digit vacant code
  - – 10-digit vacant code
  - – Nonpurchased NPA
  - – SCP overload
  - – Mass calling
  - – SMS initiated.
- A number of peg and usage counts should be provided with a recording period of 30 minutes for administration reports.   **SSP**
- Application measurements are done in 30-minute intervals, beginning on the hour (that is, in the same 30-minute intervals as the node measurements are collected). At the end of the interval, the application passes the measurements to the operations support systems.   **SSP**

Billing Measurements and Statistics

- The originating exchange (SSP) must be capable of generating AMA records for all TCAP GNS calls, using the billing information received from the SCP database. The billing indicator field is a mandatory field in the response and therefore requires that an AMA record be generated. An AMA record should be made of all answered calls. To study the service in operation, the SSP should be able to record unanswered calls.   **SSP**
- The billing data collected at the originating SSP is formatted into billing records, which are automatically transferred to a service center that generates the actual customer bills.   **SSP**

Exceptions

- The SCP must contain service logic to handle exceptions such as overload situations and database errors.   **SCP**
- The SCP node detects an overload by measuring the delays between receiving a query and returning a response to the query. When the SCP node determines overload, it should notify the application with a message. The application should respond with certain actions to help alleviate the overload. The actions taken depend on the level of overload which is indicated to the application by the SCP, see 2.11.3, "Overload Handling".   **SCP**
- Service logic must also exist in the SSP to handle other message types from the SCP, such as Automatic Call Gapping (ACG).   **SSP**

Supporting Functions

To load the SCP databases, the SMS will first produce load files. These files can be transferred via an X.25 link or by a tape transport to the SCP, to create the initial service database. The application must generate a response file, containing one response message for each message in the SMS load file. The response file is written onto magnetic tape and transferred to the SMS (via an X.25 link or tape transport).

SCP

6.2 Functional Units in the EWSD SSP

Initial Call Handling

This unit is responsible for initial GNS call handling.

Processing Functions

1. The caller dials a special access number (for example, 800) and the virtual number.
2. The SSP switch function program (call processing) in the LTG collects the digits and sends the digit block to the digit analysis program in the Coordination Processor (CP).
3. The digit analysis program in the CP (based on the trigger table) identifies that this number requires special treatment, that is, the number requires further translation by an SCP.
4. The CP assembles all destination-related information from the database, including the Destination Point Code (DPC), GTT indicator, and DPC status.
5. The Coordination Processor then sends a TCAP request to the *GNS Service Unit* in the LTG (which sent the digit block). It provides the Line Trunk Group with the Destination Point Code of the SCP to which the query will be sent, or the DPC of the STP that will perform the global title translation.

GNS Service Unit

This unit is responsible for all GNS query requests to an SCP database, and all responses from the SCP. It normally resides in the LTG.

Processing Functions

1. This unit verifies the DPC status.
2. The unit assigns the Transaction ID and the Invoke ID for the *Provide Instruction* operation. It formulates the TC component *Invoke*, and passes it to TCAP.
3. The *Invoke (Provide Instructions)* contains information on:
   - Calling number or origin
   - Called number
   - Exchange ID
   - Local Access Transport Area (LATA) ID (U.S. only).
4. The message is sent using the platform's TCAP sending capability which uses the CCS7 lower layers (SCCP class 0 and MTP) to deliver the message to the destination SCP. TCAP formulates the query. It also provides component and

transaction level processing and passes it to the SCCP. TCAP also provides the association at transaction level and the correlation at the component level.

5. The SSP establishes a three second nominal timeout counter waiting for the response.

6. In case of timeout, the SSP may send an announcement or reorder tone to the caller.

7. The response message from the SCP passes through the MTP, SCCP and TCAP before arriving at the application. The MTP and SCCP provide the transaction and deliver the message to the application. Three basic response types are expected from the SCP, namely:

Invoke (Connect Call)	To return a normal result (U.S. only)
or	
Return Result	To return a normal result (CCITT)
Return Error	To indicate an error in handling the query (normally due to an SCP error)
Reject	To reject a query due to message or format error.

8. On "Return Error" or "Reject" response, the SSP handles a call as the timeout case.

9. On normal response (Return Result or "Connect Call" Invoke), the SCP will return:
   - A translated number (C-number) or a call treatment indicator if unsuccessful in translating the number or call not allowed.
   - Automatic call gapping
   - Billing indicator (used for AMA)
   - Optional request for call termination information.

10. The SSP will use the C-number to complete the call as a normal call. This is usually done by re-initiating the translation procedure with the new C-number. The call will then proceed as a normal call.

11. For successful completion of a call, the SSP will produce an AMA record at call termination time. The billing indicator returned in the TCAP response message from the SCP is used.

12. The appropriate traffic counts are pegged by the standard call processing routines.

13. If the SCP returns an Automatic Call Gapping (ACG) response, the appropriate call gapping parameters are sent to the call gapping unit. This normally results in the SSP stopping any pending requests to the SCP for the duration of the gapping interval.

14. If the SCP so requests, the SSP sends a return result message containing call termination information.

15. If the SSP fails to complete the call as specified in the reply message from the SCP, it sends the following responses to the SCP:

Invoke (Return Error)	If the SCP did not request any call termination information.
Return Error	If the SCP requested call termination information.

16. If the SSP detects a protocol error in the response message from the SCP, it sends a "Reject" response message to the SCP assuming the SSP has found enough information to include in the Reject message.

AMA Generation
This unit is responsible for generating the appropriate AMA record.
Processing Functions
1. At call completion, the SSP produces an AMA record based on the billing indicator field from the SCP. AMA is usually handled in the CP.
2. The AMA record will contain information about the calling party, called party, billing number, call duration, answer condition, and so on.
3. The AMA records are stored on disk and later transferred to a billing center that generates the actual customer bills.

Traffic Measurements
This unit maintains all required traffic and maintenance measurements relating to GNS.
Processing Functions
1. Several peg counts and usage counts are provided with a recording period of thirty minutes. Most of these measurements apply to all services.
2. Application measurements are made twice an hour; the first starts on the hour and the second starts on the half-hour. At the end of the interval, the measurements are passed to the operations support systems.
3. The traffic measurements that are more specific to number services are:
   • Number services SCP - initiated control discrete
   • Total originating number services attempts
   • Number services calls blocked for excessive calling to vacant codes or from nonpurchased numbering plan areas
   • Number services calls blocked by SCP overload controls
   • Number services blocked for mass calling controls
   • Number services blocked by SMS-initiated controls
   • Counts for control lists overflow.

Automatic Call Gapping
This unit handles automatic call gapping requests from the SCP.
Processing Functions
1. The SCP sends the SSP a request for ACG embedded in the return message when SCP detects a certain overload condition to a particular GNS destination.
2. The ACG request is intended to reduce the traffic load to the SCP. The SCP returns the following parameters when it requests ACG:
   • Control Code
   • Control Cause Indicator
   • Control (Gap) Interval
   • Control Duration.
3. The ACG Control Code is the code affected by ACG, for example NPA, NPA NXX, NPA NXX XXXX.

4. The ACG gap interval is the time between successive releases of queries from the SSP to the SCP.
5. The ACG Control Duration is the maximum time ACG should remain in effect.
6. When an ACG control is initiated in the SSP, a timer is set to mark the duration of the control. The duration times vary from one to 2048 seconds or infinity. When the duration timer expires, the control is removed.
7. When an ACG control is initiated, a gap timer is set for a period of the gap interval. All subsequent calls to the controlled code will be blocked (with no query message sent) until the gap timer expires. The next attempt to arrive after the gap timer expires will not be blocked. This attempt will be processed in the usual way, and then the gap timer will be reset to start another blocking period. This cycle will continue until the duration timer expires. When the duration timer expires, the control will be removed without waiting for the gap timer to expire. The gap interval ranges from 0 to 300 seconds.
8. ACG can be removed through a new request from the SCP.

6.3 Administrative Units in SSP

Trigger Table Administration
Processing Functions
1. Perform initial table load.
2. Provision the trigger indicators and the destination address to which the request for GNS related queries are to be sent.
    The following parameters are required:
    • Trigger indication
    • CCS7 destination address
    • GTT indicator
    • Primary/backup.
3. Perform the following functions:
    CREATE   Make an entry in the table
    MODIFY   Edit a field(s) within a table entry
    CANCEL   Cancel a table entry
    DISPLAY   Display a table entry.
    Note: Table entries are indexed using a key.
    These man-machine commands are normally entered either from a local terminal connected to the SSP or remotely from an operations and support system center.

6.4 Functional Units in SCP

Inquiry SSP-SCP, Invoke (Provide Instructions) and SCP-Response
A service user has made a call. The SSP sends a TCAP message with the component type *Invoke (Provide Instructions)* to the SCP. The request is valid and the database query successful. The SCP application program generates a TCAP message with the component type *Invoke (Connection Control or Send Notification)* and stores the message in the output queue (the Application Platform sends the message to the SSP).
Note: The CCITT standards use *Return Result* instead of *Invoke (Connection Control)*. In the following descriptions, both component types are synonymous.

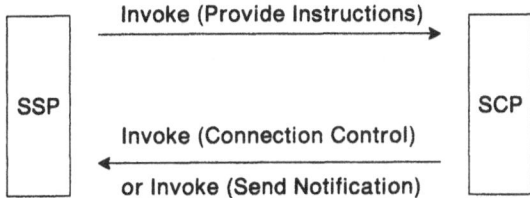

Processing Functions
1. Read the message from application platform input queue.
2. Get the main storage resources.
3. Analyze and process the header or SCCP control data of incoming message
4. Analyze and process the TCAP control information included in the message.
5. Save the header or SCCP control data and TCAP information.
6. Check and validate the input data of the message.
7. Build a command for a record search on the GNS database. The parameters of the command are file name, key (the number called), input-output area, access parameters, return code area, and so on.
8. Issue database read (the actual read of the database is performed by the application platform data manager).
9. After completion of database read, check and analyze the delivered return code.
10. Record found (no error condition).
    *   Process the data of the received record to determine the target number, considering day of week, time of day, screening, and so on.
11. Create and format the data part of the output message.
12. Restore the header or SCCP control data, and TCAP information. Determine final header or SCCP control data, and the following TCAP information:
    *   Package type identifier
    *   Responding transaction ID
    *   Component type (Invoke)
    *   Correlation ID

- Operation code (connection control no reply expected, or send notification termination)
- Digits (carrier identification)
- Billing indicators
- Network routing numbers.

13. Create an output message consisting of header or SCCP control data, TCAP control information and data part.
14. Write the output message on application platform CCS7-output queue.
15. Update the counters used for traffic measurements and statistics.
16. Write the GNS inquiry information on application journal.
17. Free main storage resources.

SSP-SCP Return Result

The SSP was able to complete the call with the returned message from the SCP after an inquiry. The return message from the SCP was defined with "Sending Termination Information". The SSP sends a message with the TCAP component type *Return Result*.

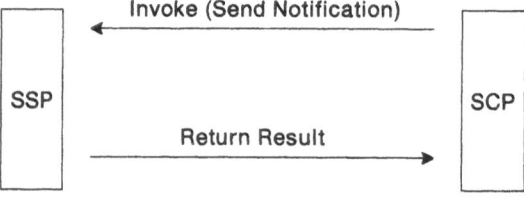

Processing Functions

1. Read the message from application platform input queue.
2. Get the main storage resources.
3. Analyze and process header or SCCP control data of incoming message.
4. Analyze and process the TCAP control information included in the message.
5. Save the header or SCCP control data and TCAP information.
6. Check and validate the input data of the message.
   - If normal end of call (no error), calculate the connect time (time elapsed between answer and disconnect).
   - If user abandon (error), calculate the connect time, create an error message, and write it to the error file.
7. Update the counters used for traffic measurements and statistics.
8. Free the main storage resources.

SSP-SCP Invoke (Report Error)
The SSP is unable to complete the call using the message returned from SCP after an inquiry. The return message from the SCP was component type "Invoke" with operation code "Connection Control" and with the definition "No Sending Termination Information". The SSP sends a message with TCAP component type *Invoke* and the operation code *Report Error*.

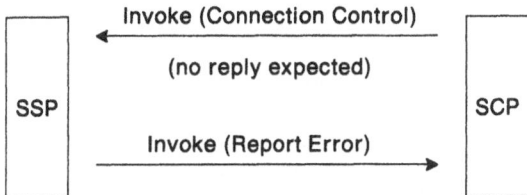

Processing Functions
1. Read the message from application platform input queue.
2. Get the main storage resources.
3. Analyze and process the header or SCCP control data of incoming message.
4. Analyze and process the TCAP control information included in the message.
5. Save the header or SCCP control data and TCAP information.
6. Check and validate the input data of the message.
    * Data not available
    * Unexpected data element value
    * Problem data.
7. Process the error data, build an error message.
8. Write the error message to the error file (the actual write is performed by the application platform data manager).
9. Depending on the error data, send a message to the GNS responsible Telco operator.
10. Update the counters used for traffic measurements and statistics.
11. Free the main storage resources.

SSP-SCP Return Error
The SSP is unable to complete the call using the message returned from the SCP after an inquiry. The return message from the SCP was "Sending Termination Information", which means TCAP component type "Invoke" with operation code "Send Notification Termination". The SSP sends a message to the SCP with TCAP component type *Return Error*.

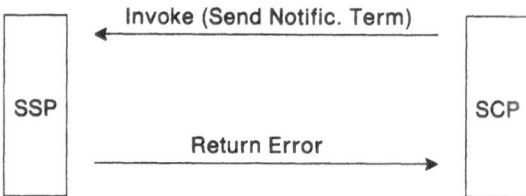

Processing Functions
1. Read the message from application platform input queue.
2. Get the main storage resources.
3. Analyze and process the header or SCCP control data of incoming message.
4. Analyze and process the TCAP control information included in the message.
5. Save the header or SCCP control data and TCAP information.
6. Check and validate the input data of the message. If:
    *   Abnormal end of call (error detected)
        −   Analyze error code
        −   Unavailable network resource
        −   Unavailable data
        −   Unexpected data value (analyze problem data)
        −   Protocol error (problem code from SSP)
        −   User error code (customer abandon).
    *   Control list overflow (a control list overflow has occurred after mass calling control)
        −   Perform actions to be specified.
7. Process the error data, build error message.
8. Write the error data to the error file (the actual write is performed by the Application Platform Data Manager).
9. Depending on the error data (for example, control list overflow), send the error message to the Telco GNS responsible operator.
10. Update the counters used for traffic measurements and statistics.
11. Free the main storage resources.

Inquiry SSP-SCP and SCP-Response, Invoke (Play Announcement)
A service user has made a call. The SSP sends a message with component type Invoke (Provide Instructions) to the SCP. The request is valid but the database query is not successful. The SCP application program generates a message with the component type *Invoke (Play Announcement)* and stores the message in the output queue (the application platform sends the message to the SSP).

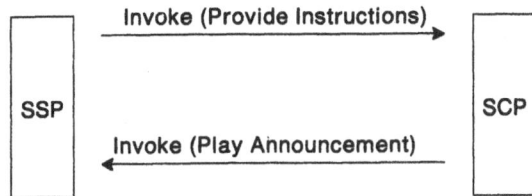

Processing Functions

1. Read the message from application platform input queue.
2. Get the main storage resources.
3. Analyze and process the header or SCCP control data of incoming message.
4. Analyze and process the TCAP control information included in the message.
5. Save the header or SCCP control data and TCAP information.
6. Check and validate the input data of the message.
7. Create a command for a record search on GNS database. The parameters for the command are file name, key (being the number called), input/output area, access parameters, return code area and so on.
8. Issue a database read (the actual read of the database is performed by the application platform data manager).
9. After completion of the database read, check and analyze the delivered return code.
10. Error condition:
    - The customer has dialed an unassigned GNS number or the GNS number was outside of the band.
11. Create and format the data part of the output message.
12. Restore header/SCCP control data, and TCAP information. Determine final header/SCCP control data, and the following TCAP information.
    - Component type (Invoke)
    - SSP Invoke ID
    - Correlation ID
    - Operation code (play announcement)
    - Standard announcement, and so on.
13. Build an output message consisting of header/SCCP control data, TCAP control information and data part.
14. Write the output message on application platform CCS7-output queue.
15. Update the counters used for traffic measurements and statistics.
16. Write the GNS inquiry information on the application journal.
17. Free the main storage resources.

Inquiry SSP-SCP and Response SCP-SSP with Return Error

A service user has made a call. The SSP sends a message with the component type Invoke (Provide Instructions) to the SCP. The request is valid but the SCP is unable to provide routing instructions, because of improper or invalid data in a component of the query message. The SCP generates a message with component type *Return Error*.

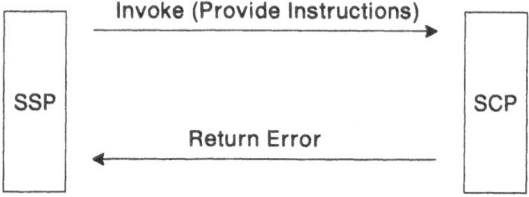

Processing Functions

1. Read the message from application platform input queue.
2. Get the main storage resources.
3. Analyze and process the header/SCCP control data of incoming message.
4. Analyze and process the TCAP control information included in the message.
5. Save header/SCCP control data and TCAP information.
6. Check and validate the input data of the message.
7. Invalid or improper data in a component is detected.
8. Create an error message, write the message to the error file.
9. Create and format the data part of the output message.
10. Restore the header or SCCP control data, and TCAP information. Determine final header or SCCP control data, and the following TCAP information.
    *   Package type identifier
    *   Responding transaction ID
    *   Component type (Return Error)
    *   Correlation ID
    *   Error code
    *   Problem data, if any.
11. Build an output message consisting of header/SCCP control data, TCAP control information and data part.
12. Write the output message on application platform CCS7 output queue.
13. Update the counters used for traffic measurements and statistics.
14. Free the main storage resources.

Inquiry SSP-SCP and Response SCP-SSP because of Protocol Error

The SSP sends a message with a protocol error (probably an application error in the SSP). The SCP detects the error, writes an error record, informs the Telco operator, and sends a message with component type "Reject" to the SSP.

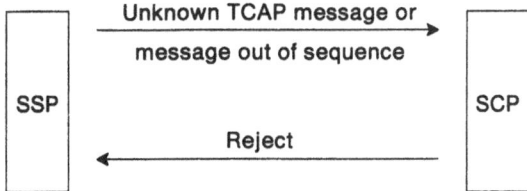

Processing Functions
1. Read the message from application platform input queue.
2. Get the main storage resources.
3. Analyze and process the header/SCCP control data of incoming message.
4. Analyze and process the TCAP control information included in the message.
5. Save the header or SCCP control data and TCAP information.
6. A protocol error is detected.
7. Create an error message, write the message to the error file.
8. Create an operator message and send the message to the Telco responsible GNS operator.
9. Restore the header or SCCP control data, and TCAP information if possible. Determine final header or SCCP control data, and the following TCAP information:
   - Package type identifier
   - Responding transaction ID
   - Component type (Reject)
   - Correlation ID
   - Problem code.
10. Create an output message consisting of header or SCCP control data, TCAP control information, and problem code.
11. Write the output message on application platform CCS7-output queue.
12. Update the counters used for traffic measurements and statistics.
13. Free the main storage resources.

SSP-SCP Reject
The SSP has found a protocol error in a response message of the SCP and enough information to include in the reject message, for example, the responding transaction ID, or the correlation identification. The SSP sends a message to the SCP with TCAP component type *Reject*.

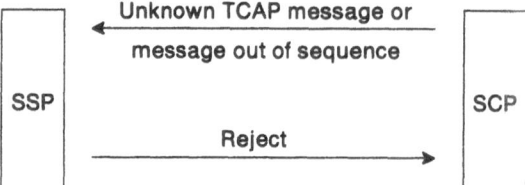

Processing Functions
1. Read the message from application platform input queue.
2. Get the main storage resources.
3. Analyze and process the header/SCCP control data of incoming message.
4. Analyze and process the TCAP control information included in the message.
5. Save the header or SCCP control data and TCAP information.
6. Check and validate the input data of the message.
7. A protocol error has occurred.
8. Analyze the error code.
9. Process the error data, and build an error message.
10. Write the error data to the error file (the actual write is performed by the application platform data manager).
11. Build an operator message, and send it to the Telco responsible GNS operator.
12. Update the counters used for traffic measurements and statistics.
13. Free the main storage resources.

SCP-SSP Automatic Call Gapping
An asynchronous CICS transaction is started periodically to measure the load of the application. The parameters for the load measurement are the response time of the transactions in a given time frame, the amount of the traffic counters, the database load, a destination mass calling or unprocessed application requests. The ACG can be initiated by an overload situation, or externally by an operator message.

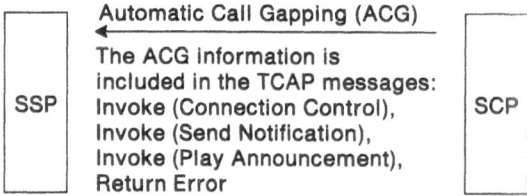

Processing Functions
1. Store the traffic counters of the last measurement period and determine the amount of transactions in this period.
2. Analyze the database load in this time period.
3. Analyze the application input queue for unprocessed application requests.

4. Analyze the calls for destination mass calling.
5. Analyze the response time counters.
6. Check the number of active application clone tasks.
7. Process the parameters (database load, input queue length, and response time counters) to determine the load.
   - Normal load situation: No action
   - Overload situation, or ACG initiated by an operator: The overload level and sublevel will be obtained by a table. Input values for this table are the above mentioned parameters, or the value of the operator message. Output is the level and sublevel of the overload situation.
8. Create output message consisting of header/SCCP control data, and the following TCAP ACG information.
   - Component type (Invoke)
   - Operation code (ACG)
   - Digits (called party address)
   - Automatic Call Gapping indicators.
9. Write output message on Application Platform CCS7-output queue.
10. Update counters for ACG control and statistics.
11. Write information on application journal.
12. Free the main storage resources.

6.5 Administrative Units in SCP

SMS - SCP Retrieve Application Measurements for GNS
This command retrieves the application measurement for the GNS application, which maintains measurement data. The SMS polls the SCP-node on a fixed time basis for measurement data. The SCP is able to generate several types of reports: daily, hourly, and every five minutes. The name of the reports, schedules, and so on, are maintained in a parameter control file.

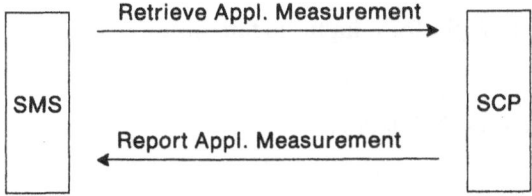

Processing Functions
1. Read the message from application platform input queue.
2. Get the main storage resources.
3. Analyze and process the message data.

4. Save the message.
5. Read the application measurement counters.
6. Using the measurement counters, create the following application report message:
    - Type—application identification
    - More—indicates that data is sent for one or more intervals
    - Time Interval—begin and end time of the interval
    - Data, number of:
        - Successful responses
        - Response messages with errors
        - Misrouted queries
        - Queries rejected.
7. Write an output message on application platform SMS output queue.
8. Free the main storage resources.

SMS - SCP Update Processing of GNS Database

An update is an addition, change or deletion to a data record. It can be a single transaction or a batch of transactions affecting many data records. The update is a logical unit of work; if an abnormal termination or failure occurs, all the changes of a logical unit of work are backed out, either by dynamic transaction backout or emergency restart.

The following process assumes that if an SMS update is not acknowledged within a prescribed period of time, the SMS will resend the update.

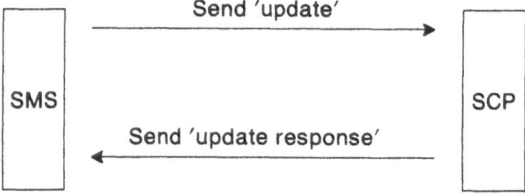

Processing Functions

1. Read the message from the application platform input update queue.
2. Get the main storage resources.
3. Analyze, check, and edit the message data.
4. Create command for record update on GNS database.
5. If:
    - a. Update unsuccessful:
        - Update request was to change or delete a record.
            - Record not found condition.
            - Sequence number not valid.

- Time stamp too "old".
  Write the update request to the error file. Create the response "update not successful".
- Update request was to add a new record.
  - Record exists already.
    Write the update record to the error file. Create the response "update not successful".

b. Update successful:
- The platform data manager:
  - Saves the previous version of the changed record.
  - Marks the data file in process for updating.
  - Sends the changed record (or delete or add request) to the update task in the subordinate hosts.
  - The update task changes the record and sends the "update successful" status.
- Create the response "update successful".

6. Write the response message to the SMS output queue.
7. Write the GNS update record on application journal.
8. Free the main storage resources.

GNS Database Customizing and Maintenance
The database is organized as a Virtual Storage Access Method (VSAM) Key Sequential Data Storage (KSDS). This allows "direct" access to records via primary or secondary keys. VSAM KSDS supports variable length records, reusable space and buffer pooling.

Processing Functions
1. Database description
   - Record layout
     - Primary key
     - Secondary keys
     - Data fields
   - Define catalog
     - Master catalog
     - User catalog
   - Define cluster
     - Index and data cluster
       - Control interval size (CI)
       - Control area size (CA)
       - Free space
       - Buffer space
       - Other parameters
2. Database initial load
   The load file is sent from the SMS to the SCP via a file transfer program. The SCP receives the data and the new records are stored to disk. The data is defined as logical units of work, that means there are possible restart points in the case of an abnormal end. The records are stored to a disk via a VSAM

function which creates the index and data areas with the defined CI/CA sizes and free space. The SCP application produces a response file containing the amount of successful "loaded" records and a message for each "unsuccessful loaded" record. The response file is sent back from the SCP to the SMS via the file transfer program. A similar program is required for bulk inserts.

3. Database maintenance and reorganization
   * Standard batch utilities are used.
       − Display of statistics in the VSAM catalog gives information about the need of reorganization. The different parameter values giving rise to a reorganization must be defined.
       − During reorganization, the database is not available for queries and update requests, the Customer Information Control System (CICS) is shut down. The query and update requests are performed by other SCP components (other CICS systems) with "duplicated" databases. The CICS is started, after the reorganization. A VTAM session with all other SCP components is established. Updates made on another SCP during reorganization time are worked into the reorganized database. All "duplicated" databases are to be reorganized with this procedure.

GNS Initialization Routine in the Platform (CICS) Start Up Phase
Processing Functions
1. In the start up phase each application must be initialized.
2. The routine must create the input, output and save queues for the application, and a Common Work Area (CWA). The CWA is needed for queue pointers, counters and threshold parameter values. The routine must set the counters, threshold and data fields to initial values.
3. The initialization routine loads all application relevant tables, such as the country-specific application error message file. The routine must open all application relevant data files.

GNS Application Error Message File
Processing Functions
1. When an application program detects errors, it sends messages to the operator or to system components. The target of the messages is defined in a "message filter table" as part of the message file. The text for all application programs and the filter table are stored in the error message file.
2. Messages are accessed via a message number thus making it easy to change the message layout without changing the application programs. Furthermore, it facilitates national language support (for example, in the start up phase, when the country-specific message file is to be loaded).

Part 3. Alternate Billing Service

This part provides detailed information about the Alternate Billing Service. It examines what the service is, how it operates, and the factors to consider when implementing ABS.

This part is ABS-specific; it adds to, and does not replace "Part 1. Overview of the Intelligent Network" of this report.

Chapter 7. ABS Service Description

7.1 Overview

Basic Service

With the Alternate Billing Service (ABS), a user can bill a call to a number other than the calling (A) number, or to a Telco-provided calling card.

In the United States an ABS call requires a special prefix (for example, 0) and gives the caller the option to bill the call to a calling card account, a third party number, or the called (B) number.

It triggers the interrogation of a database to validate the billing of the call. The database validates the calling card account, handles third party billing, or handles called party billing (collect call).

The system might prompt the caller for additional digits (for example, the calling card number) depending on the option selected.

Enhanced Service

An enhanced version of the ABS is an automated service so that operator services are not required for called number and third party billing.

Background

ABS is designed to replace manual operator services, however, Bellcore recommendations in the U.S. still require manual operator services for third party and collect call billing. Typical applications include calls from a payphone or a travelling salesman calling home.

Benefits

The potential benefits of ABS to the service user are:
- Convenient calling (for example, no money required)
- Region wide dialing capability
- Faster response times
- No human intervention
- Cheaper than operator assisted calls
- Single telephone bill.

The potential benefits to the service subscriber are:
- Same as for the service user

- Accounting control for business customers.

The potential benefits to the network operator are:

- Increased revenue through increased volume
- A visible service for customers
- Reduced operator costs.

7.2 Functional Description

General

ABS requires the services of a centralized database and an operator services system (OSS). In the U.S., the Line Information Database (LIDB) supports ABS. The LIDB database resides in an SCP. Both the originating SSP and the OSS have access to the LIDB through the CCS7 network. All calls requiring ABS are forwarded to an SSP office that can access the LIDB.

The Service Management System (SMS) administers the centralized database in the SCP. The SMS can accept service order information automatically (that is, scheduled) or receive updates through terminal access. For ABS, only the network operator is allowed direct access to the SMS.

The SCP contains the real time LIDB that supports ABS. It responds to queries from both the SSP and OSS. The LIDB application in the SCP performs the following functions:

- Validates calling card account and PIN digits
- Provides information about the originating line's ability to make calling card calls, collect calls, or third party calls
- Provides information about the terminating line's ability to allow, disallow or require approval for collect calls
- Provides information about the ability to allow, disallow or require approval for third party billing to a particular number.

Service User's Perspective

The service user makes a phone call intending to bill the call to an account different from the calling number. To bill a call to an alternate number the user can dial a prefix, and then the called (B) number. The system prompts the caller by a tone or an announcement, instructing the service user to enter a calling card number, enter a prefix to get an operator, or wait for an operator.

With calling card billing, the user enters his calling card number (which normally consists of a telephone number and PIN digits) using a DTMF phone. The SSP validates the calling card number by sending a query using the CCS7 network to an SCP which has the LIDB. Once validated, the system completes the call to the appropriate destination.

The user that automatically bills his call to a calling card account can bill a subsequent call to the same account, without reentering the calling card number. This is known as "sequence dialing". Within a specified period at the completion of the first call, the user is required to enter a # digit or a hook flash. The system gives him a dial tone. He is only required to enter the new called number, thus avoiding an additional query to the SCP LIDB.

Based on existing Bellcore recommendations, third party and collect call billing require the intervention of the operator services system (OSS). The operator manually collects and validates the billing instructions, using the LIDB. After proper billing validation, the operator requests completion of the call.

If the called party's approval is required for collect call billing, the operator establishes a call to the called party (B) while the calling party (A) is still on the line; verbal approval must be obtained before allowing both parties to commence their conversation.

Where third party billing approval is required, the operator establishes a call to the third party and request billing approval. Normally, the calling party is not allowed to listen to the conversation. Once approval is obtained, the operator completes the call.

Service Subscriber's Perspective

The service subscriber is billed for the phone call by a calling card account, a collect call or third party billing. The service subscriber is not allowed direct access to the database through the SMS; any administrative changes must be requested through the network operator. Accordingly, the service subscriber can:

- Obtain a calling card from the company serving his local area. The card allows him to call from anywhere in the network, and bill the call to that particular account. If the subscriber has a telephone account, calls billed to the card account appear on the subscriber's normal telephone bill. In the U.S., holders of Regional Bell Operating Company (RBOC) calling cards must have a telephone number; this is a question of the individual Telco's policies.
- The service subscriber (in this case any telephone subscriber) can request the network operator to always allow, disallow, or require approval of third party calls to his phone number, or of collect calls.
- Verbal approval for collect call or third party billing is performed on a per call basis. During the call setup, the operator will call the billed subscriber to obtain verbal approval for the charge; the subscriber can deny or accept the charges.

Network Operator's Perspective

The network operator is required to:

- Establish, on request, a calling card account for the service subscriber. This is done through the SMS, which updates the SCP database with the appropriate account information.
- Update each network subscriber's line database record (regarding information on whether they allow, disallow or require approval of third party and collect call billing to their phone numbers). This database update must be done using the SMS.
- Provide operator services (or equivalent) to collect billing information and perform validation for third party and collect call billing.

Relationship to ISDN Supplementary Services
With "reverse charging", one of the ISDN supplementary services, called number billing takes place. This requires that both parties be ISDN subscribers.

7.3 Standards

Service
The service standard and ABS TCAP messages are based on Bellcore technical recommendations.

Trunk Signaling
Normal inter-exchange trunk signaling is required for ABS. However, it is desirable that the network has the capability to automatically identify the caller by transfer of the ANI digits.

7.4 Service Interaction

Service User
1. The user dials the prefix and called number.
2. He receives a tone and/or announcement and is requested to enter his calling card number, or enter another prefix to get the operator.
3. For calling card billing, (DTMF phone) the user enters a calling card number and PIN digits.
4. The system verifies the card number and completes the call.
5. For operator assisted calls, the user provides verbal instructions to the operator, collect call, third party call, or calling card number.
   Normally, he is requested to indicate the type of billing requested, but may have to give his name, in case the operator requires verification of the request.
6. For sequence dialing, the user is required to enter either a # digit or a hook flash within a timeout period. On receipt of a new dial tone, he dials the new number without reentering the calling card number. Sequence dialing applies to calls that are automatically billed to a calling card account.

Depending on whether the originating office supplies the ANI information, the user might be requested to verbally supply his calling number. It is desirable that the network provides automatic number identification (ANI) capability.

Service Subscriber

The service subscriber is not allowed direct access to the ABS database. All requests must pass through the network operator.

The service subscriber requires no guidance. All data to be supplied by the service subscriber must be given verbally or through a written request to the network operator. The type of data requested includes:

- Name
- Address
- Telephone number
- Third party billing always allowed, disallowed or per-call approval required
- Collect call always allowed, disallowed or per-call approval required
- PIN (optional).

Network Operator

Besides the general data as outlined in 2.5.4, "Network Operator", the network operator must allow for a personal calling card number for a service subscriber. The calling card number may consist of the telephone number and PIN, which is administered by the network operator.

Access Instrument

The following applies to the service user:

- A Rotary POTS telephone set is sufficient for collect call, third party billing, and operator assisted calling card billing.
- A telephone with DTMF capability, is required for automatic calling card billing.
- Pay phones with DTMF capability allow automatic calling card billing. Otherwise, manual operator intervention is required for all billing requests.
- ISDN access is expected to provide additional capabilities, such as collect call and third party billing, without operator intervention.

7.5 Billing

SSP

The originating SSP generates an AMA record for all calls that were automatically billed to a calling card account. The AMA record includes information about the calling card account number, any subaccount number, the calling number, number called, IC used (U.S. specific), time of connection and length of the call.

OSS

The OSS generates an AMA record by recording the information listed above and operator work time. Bills should indicate whether the call received operator or automated treatment.

SCP

The SCP does not record billing information, but does provide for verification of the requested billing.

7.6 Service Logic

7.6.1 Distribution

SSP

The SSP contains originating triggers that identify the call's need to access SCP data and verify the billing function requested. The SSP should also contain logic to forward the call to the operator services system, in case operator assistance is required.

SCP

The SCP contains logic to interrogate the appropriate database entry in the LIDB of the billed party, in order to verify the requested billing function. The SCP performs the necessary verifications depending on the query type as follows:

Query to validate Calling Card number

- PIN match
- Service denial indicators
- Calling card subaccount number
- PIN restriction indicator
- IC indicators and ICs
- True billing number (optional).

Query to determine processing of Collect and Third Number calls

- Collect acceptance indicator
- Third number acceptance indicator
- Treatment Indicator
- Service or equipment indicator
- IC indicators and ICs.

LIDB description

The main database that supports ABS is the Line Information Database (LIDB) which contains line-oriented and special billing information. It is accessed for:

- Originating screening
- Billed number screening (required by ABS)
- Calling card validation (required by ABS)
- Terminating screening
- Basic intercept
- Packet Switched Public Data Network (PSPDN) support.

In the United States, the LIDB is organized according to:
- Group Records
  - Primary key is 6-digit NPA-NXX or RAO-0/1XX (where RAO means Revenue Accounting Office)
  - Data Elements (preliminary information)
    - Status
    - Processing indicators
    - Owning Telco's ID
    - Administrative system ID.
- Line Records
  - Primary key is 10-digit NPA-NXX-XXXX
  - Data Elements (preliminary information)
    - Originating calling card indicator
    - Treatment indicator
    - PIN
    - PIN restriction indicator
    - Calling card subaccount number
    - PIN service denial indicator
    - Primary preferred IC indicator
    - Primary preferred IC.
- Special billing number records.

Triggers
The SSP trigger is the prefix. For example, 0 in the United States.

7.6.2 Functional Flow

SSP/OSS - SCP
Several types of TCAP messages flow between the SSP or the OSS, and the SCP. The OSS is required for calls billed to the number called, a third party, or manually requested calling card billing. OSS has a CCS7 interface to the SCP, similar to the SSP. TCAP queries to the LIDB are generally referred to as Operation Protocol Data Units (OPDU). For normal call completion, a query (Invoke) OPDU is sent to the SCP/LIDB and a return result message is sent back with the requested information. The following are descriptions of the various messages that flow between the SSP and SCP. Figure 31 and Figure 32 show the ABS message flow.

SSP/OSS to SCP

Invoke Originating Line Number Screening (OLNS) OPDU: Obtain originating line number and screening information.

Invoke Connection Control (CC) OPDU: Obtain credit card verification.

Invoke Billed Number Screening (BNS) OPDU: Obtain billed number screening and information.

SCP to SSP/OSS

Return Result OPDU: Return a normal result.

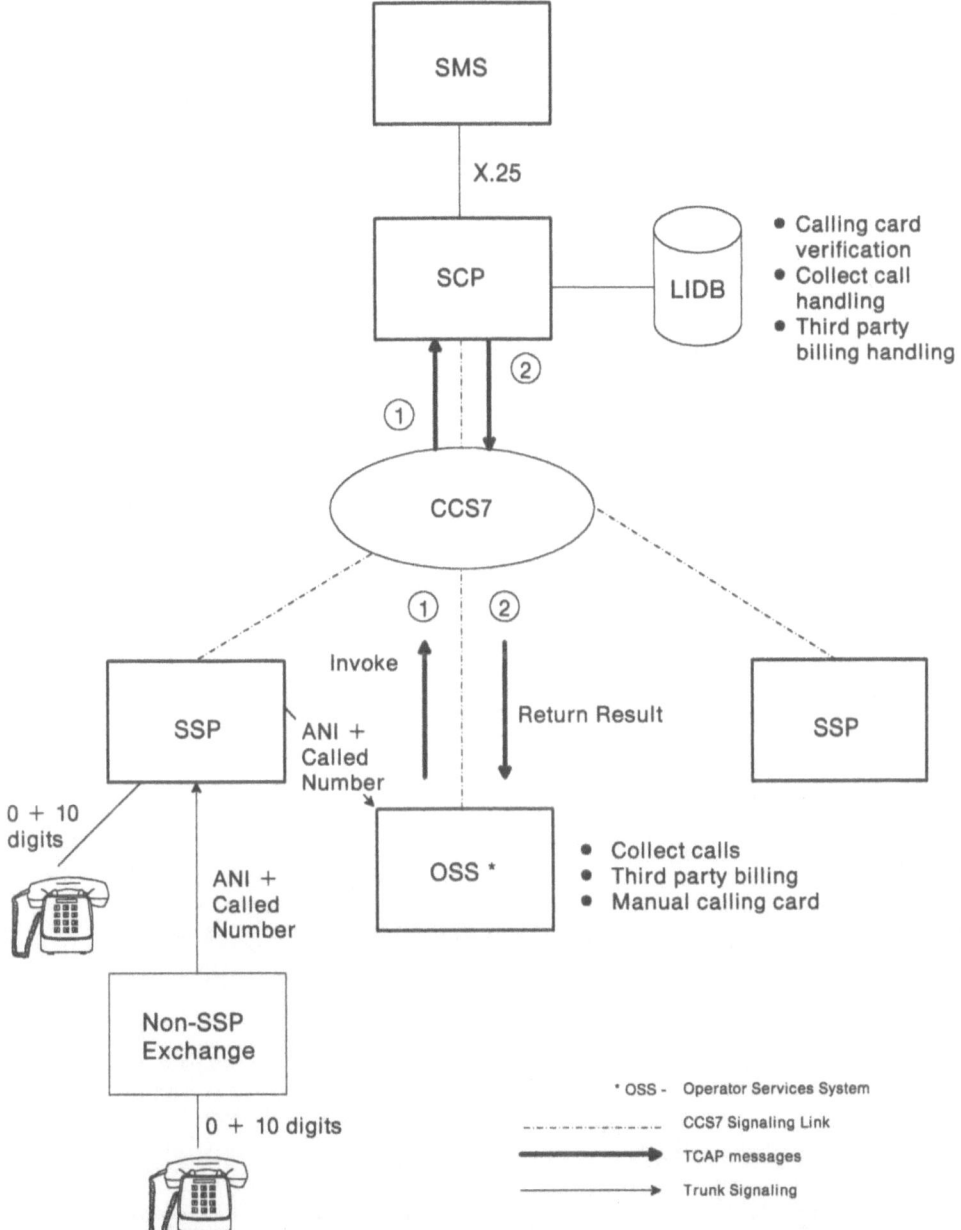

Figure 31. Functional Flow of ABS with Collect Call, Third Party Billing, and Manual Calling Card Billing

Return Error OPDU: Indicate an error in handling the query (normally due to an SCP error).

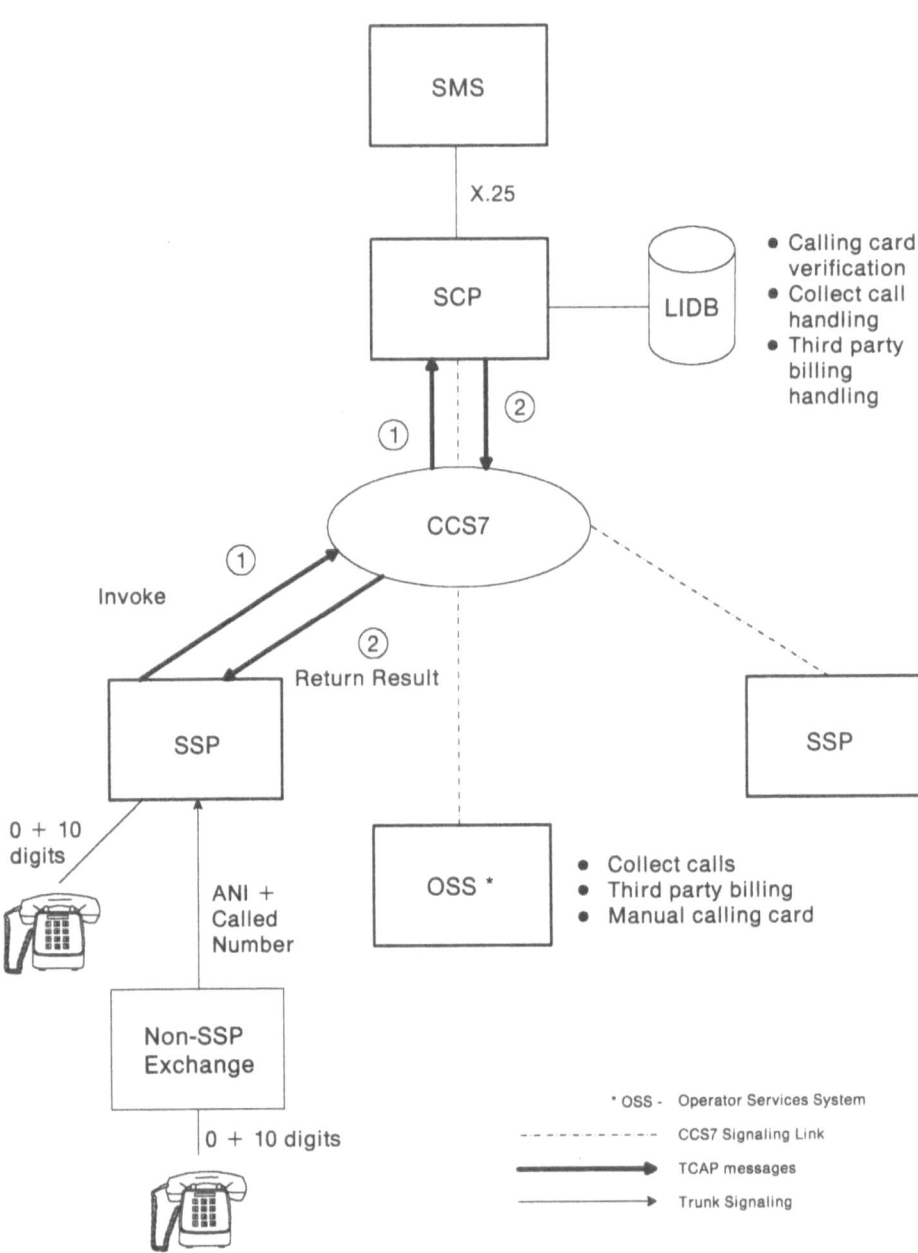

Figure 32. Functional Flow of ABS with Automatic Calling Card Billing

Reject OPDU: Reject a query due to message error or format.

The Invoke OPDU may contain:

- Calling number (12 digits)
- Called number (10 digits)
- Billing number (10 digits)
- PIN (4 digits)
- Carrier ID (3 digits U.S. specific)
- Closed user group index (2 digits).

SCP -SMS

Four classes of messages flow between the SCP and SMS:

- LIDB line and group data maintenance
- LIDB control data maintenance
- LIDB usage data administration
- LIDB status and exception reporting.

A complete list and description of these messages is available in SMS/LIDB-LIDB Interface Specification, Bellcore TA-TSY-000446.

7.7 Traffic Measurements Requirements

SSP

The SSP maintains numerous traffic measurement counts, most of which are for general capabilities. Some representative counts are suggested below:

- Total number of ABS calls
- Number of completed ABS calls
- Number of calling card calls
- Number of ABS calls requiring an OSS.

OSS

- Total number of ABS calls
- Number of completed ABS calls
- Number of collect calls
- Number of third party calls
- Number of calling card calls (requiring OSS)
- Total time required by an operator (for a specified period).

SCP

Suggested specific traffic measurements in the SCP are:

- Total number of ABS queries
- Number of calling card queries
- Number of collect call queries
- Number of third party queries.

7.8 Dynamic Requirements and Performance

SSP

The SSP should begin timing for a response from the SCP after sending a query message. A nominal value of three seconds should be used. In case of timeout, the SSP should forward the call to the OSS for further handling of the call. The OSS operator gives the caller additional options for billing the call.

SCP

Upon receipt of a query from the SSP, the SCP should respond within 1.0 second on average, and within 1.5 seconds 99% of the time.

7.9 National Dependencies

The following characteristics must be adjusted from country to country:
- Availability of operator services for called party and third party billing
- Availability of prefix
- Availability of AMA recording capability (this is important for third party billing).

In the United States, operator assisted services are currently required to handle third party and called party billing. However, Telcos in other countries may have other methods of handling these types of billing (for example, without operator intervention).

7.10 Future Considerations

- The major goal is to completely automate the ABS service, so that operator services are not required for called number and third party billing.
  The automation may involve giving the service user an option to enter the billing choice (such as collect call or third party billing, and additional digits if required). Where the LIDB indicator requires third party approval for billing, the system dials the billed number, and prompts with a recorded announcement. The third party will be asked to approve or disapprove the billing by entering a choice of digits. If approved, the system completes the call.
  ISDN already includes called party billing as a supplementary service (reverse charging). This means that called party billing will be completely automated by ISDN. The compatibility of ISDN reverse charging and IN collect calling requires further investigation.

- In the future, the acceptance of collect call and third party billing may depend on the origin of the call. This could be based on a region (for example, NPA), on a district (for example, NPA-NXX) or on a particular number (for example, NPA-NXX-XXXX).

Chapter 8. ABS Application Description

The description in this chapter is based on IBM's and Siemens' views of:
- Functional requirements and allocation. Requirements are specified and each item is assigned to the SCP, the SSP, or the OSS, depending on where it must be developed.
- Functional units. Processing functions needed to handle the requested tasks are described and grouped logically into functional units for the separate components.
- Administrative Units. Processing functions not directly required in the real time call handling process are combined into administrative units, separately for SSP and SCP.

A bold printed SSP means that the associated function must be performed in the Service Switching Point.

A bold printed SCP means that the associated function must be performed in the Service Control Point.

A bold printed OSS means that the associated function must be performed in the Operations Support System.

8.1 Functional Requirements and Allocation

Real Time Call Handling
1. The service user dials a prefix and a number. He receives a tone or announcement and is requested to enter calling card number or use another prefix to get the operator.   **SSP**
2. For calling card billing, the user enters the calling card number and personal identification number (PIN) digits. This data is collected by the SSP.   **SSP**
3. The SSP should contain originating triggers that identify the need to access the SCP data to verify the billing function requested. The trigger for Alternate Billing Service (ABS) is the prefix (for example, 0).   **SSP**
4. The SSP sends a query (TCAP message) using the CCS7 network to an SCP which has the LIDB database.   **SSP**
5. For operator assisted calls, the SSP should contain logic to forward the call to the Operator Services System (OSS).   **SSP**

6. The OSS sends a query (TCAP message) using the CCS7 network to an SCP
    with the LIDB database.   **OSS**
7. The SCP contains the real time LIDB supporting ABS. It responds to queries
    from the SSP and OSS. The LIDB application in the SCP performs the
    following functions: **SCP**
    - Validates the calling card account and PIN digits:
      - PIN match
      - Service denial indicators
      - Calling card subaccount number
      - PIN restriction indicator
      - Interexchange carriers
      - True billing number (optional).
    - Provides information about originating line: its ability to make calling card
      calls, collect calls, and third party calls.
    - Provides information about terminating line: its ability to allow, disallow,
      or require approval for collect calls.
    - Provides information about third party billing; ability to allow, disallow or
      require approval for third party billing to a particular number:
      - Collect acceptance indicator
      - Third number acceptance indicator
      - Treatment Indicator
      - Service or equipment indicator
      - IC indicators and ICs.
8. After validation of the calling card number or billing information, the SSP or
    OSS completes the call.   **SSP**
9. Users who automatically bill their call to a calling card account can, on
    completion of a call, bill a subsequent call to the same account without
    reentering the calling card number. This is known as sequence dialing. The user
    is required, within a limited time after the completion of the first call, to enter a
    # digit or a hook flash. The system gives a dial tone, and the user needs only to
    enter the new number. He is billed using the previously entered calling card
    number, thereby avoiding an additional query to the SCP LIDB.   **SSP**
10. The SSP should begin timing for a response from the SCP after sending a query
    message. A nominal value of three seconds should be used. In case of timeout,
    the SSP should forward the call to the OSS for further handling of the call. The
    OSS operator will give the caller additional options for billing the call. **SSP**
11. Several types of TCAP messages must flow between the SSP or OSS and SCP.
    TCAP queries to the LIDB database are generally referred to as Operation
    Protocol Data Units (OPDUs). For a normal call, a query (Invoke) OPDU is
    sent to the SCP/LIDB and a Return Result message is sent back with the
    requested information. The following are descriptions of the various messages
    that flow between the SSP and SCP:
    - **SSP/OSS to SCP**
      - Invoke OLNS OPDU—Obtain originating line number screening
        information
      - Invoke CC OPDU—Obtain credit card verification

- Invoke BNS OPDU—Obtain billed number screening information.
- **SCP to SSP/OSS**
  - Return Result OPDU—Return a normal result
  - Return Error OPDU—Indicate an error in handling the query (normally due to an SCP error)
  - Reject OPDU—Reject a query due to message error or format.
12. The Invoke OPDU may contain:
  - Calling number - 10 digits
  - Number called - 10 digits
  - Billing number - 10 digits
  - PIN - four digits
  - Carrier ID - three digits
  - Closed User Group index - two digits.

SMS-SCP Transactions
1. Update the SCP database with calling card account information.   **SCP**
2. Update for each of the network subscriber's line database records, considering whether they would always allow, always disallow or require approval of third party and collect call billing to their phone numbers.   **SCP**
3. Four classes of LIDB messages flow between the SCP and SMS: **SCP**
  - Line and group data maintenance
  - Control data maintenance
  - Usage data administration
  - Status and exception reporting.

Traffic Measurements
1. It is anticipated that ABS specific counts will take place in the SSP. These counts include: **SSP**
  - Total number of ABS calls
  - Number of completed ABS calls
  - Number of calling card calls
  - Number of ABS calls requiring an OSS.
2. It is anticipated that ABS specific counts will take place in the SCP. These counts include: **SCP**
  - Total number of ABS queries
  - Number of calling card queries
  - Number of collect call queries
  - Number of third party queries.
3. A number of peg and usage counts should be provided. Administration reports should be recorded every 30 minutes.   **SSP**
4. Application measurements are made in 30-minute intervals as the node measurements are collected, beginning on the hour: At the end of the interval the application passes the measurements to the operations support system. **SSP**

Billing Measurements and Statistics
The originating SSP generates an AMA record for all calls that were automatically
billed to a calling card account. The AMA record includes information about the
calling card account number, any subaccount number, calling number, number
called, IC used (U.S. only), connect time and length of call.   **SSP**

Exceptions
The SCP node detects an overload by measuring the delay between receiving a
query and returning a response. When the SCP node determines an overload, it
notifies the application with a message and the application responds with certain
actions to help alleviate the overload. The action taken depends on the level of
overload which is indicated to the application by the SCP. See 2.11.3, "Overload
Handling".   **SCP**

Supporting Functions
The SMS produces load files when loading the SCP databases. These files can be
transferred to the SCP via an X.25 link or by tape transport, to create the initial
service database. The application must generate a response file, containing one
response message for each message in the SMS load file. The response file is written
onto magnetic tape and transferred to the SMS (via an X.25 link or tape transport).
                                                                                 SCP

8.2 Functional Units in SSP

Initial Call Handling
This unit handles all ABS calls requiring operator assistance, or calling card billing
requiring SCP verification.

Processing Functions
1. The user dials a prefix and waits for an operator, or dials the prefix and a
   number.
2. If:
   a. The user dials a prefix only, the SSP connects the user to an operator
      services system for call handling. No additional activity is expected in the
      SSP.
   b. The user dials a prefix and a number, the call processing program in the
      LTG collects digits and sends the digit block to the analysis program in the
      CP.
3. The CP determines that additional calling card digits and PIN are required. It
   connects the caller with a tone or an announcement, instructing the caller to
   enter the calling card number or a prefix for operator assistance.
4. The CP then connects the caller to a DTMF receiver, to collect the information
   that the caller has entered.
5. If the caller enters another prefix or times out, the SSP connects him to the OSS
   for manual handling of the call.

6. If the caller enters calling card digits, the SSP collects the account number and PIN digits.
7. The CP digit analysis program determines (based on the trigger table) whether verification of the calling card account from the SCP is required. If verification is required, the CP assembles all the destination related information from the database, including the DPC, GTT indicator, and DPC status.
8. The CP sends a TCAP request to the ABS Service Unit in the LTG (which sent the digit block). It provides the LTG with the DPC of the SCP to which the query is sent or the DPC of the STP that will perform the global title translation.

ABS Service Unit
This unit normally resides in the LTG, and handles ABS requests for billing verification from the SCP, and processes all TCAP responses from the SCP.

Processing Functions
1. Verifies the DPC status.
2. Assigns the transaction ID and invoke ID for the Invoke OPDU operation, and formulates the TC component Invoke and passes it to TCAP. The Invoke CC OPDU is used for calling card verification.
3. The Invoke OPDU contains:
   - Calling number - 10 digits
   - Number called - 10 digits
   - Billing number - 10 digits
   - PIN - four digits
   - Carrier ID - three digits (U.S. only)
   - Closed User Group index - two digits.
4. The message is sent using the platform's TCAP sending capability, which uses the CCS7 lower layers (SCCP class 0 and MTP) to deliver the message to the SCP. TCAP formulates the query, provides component and transaction level processing, and passes it to the SCCP. TCAP also associates the transaction with the appropriate reply at the transaction level and the correlation at the component level.
5. The SSP waits three seconds for a response before timing out.
6. In case of timeout, the SSP may send an announcement or reorder to the caller, or connect him to an operator (OSS).
7. The response message from the SCP passes through the MTP, SCCP and TCAP before arriving at the application. MTP and SCCP provide appropriate processing at their levels. TCAP terminates the transaction and delivers the message to the application. Three basic response types are expected from the SCP, namely:

Return Result OPDU	Used to return a normal result
Return Error OPDU	Used to indicate an error in handling the query (normally due to an SCP error)
Reject OPDU	Used to reject a query due to message or format error

8.   On receiving a "Return Error" or "Reject" OPDU response, the SSP handles the call as if it were a timeout case.
9.   On receiving a normal response (return result), the SSP completes the call (if the calling card account was verified) or rejects the call (if the SCP failed to verify the account).
10. For successful call completion, the SSP produces an AMA record at call termination time. The AMA record contains information about the calling party, called party, billing number, call duration, and so on.
11. The appropriate traffic counts are pegged by the standard call processing routines.
12. If the SSP detects a protocol error in the response message from the SCP, it sends a "Reject" response message to the SCP. This assumes that the SSP has found enough information to include in the reject message.

Sequence Dialing Handling
This function handles the sequence dialing service feature.

Processing Functions
1.   Sequence dialing allows the user to begin a new call and bill it to the calling card account number previously entered.
2.   At the end of a call that was automatically billed to a calling card account, the SSP establishes a timeout period within which the user can originate a new call.
3.   The SSP awaits a # DTMF digit by connecting a DTMF receiver on the line, or monitors for flashhook signal.
4.   If either of these signals are received within the allowed time, the user can establish a new call without reentering a calling card number.
5.   The call is billed to the same account number that was previously entered, and does not require an SCP query.

AMA Generation
This unit is responsible for generating the appropriate AMA record.

Processing Functions
1.   At the end of a call, the SSP produces an AMA record based on the billing indicator field from the SCP. AMA is usually handled in the CP.
2.   The AMA record contains information about the calling card account number, subaccount number (if any), calling number, number called, PIN, IC used (US only), connect time and length of call.
3.   The AMA records are stored on disk and later transferred to a billing center that generates the actual customer bills.

Traffic Measurements
This function maintains all ABS traffic and maintenance measurements.

Processing Functions
1.   Peg counts and usage counts are provided with a recording period of thirty minutes.
2.   Application measurements are made in 30-minute intervals on the hour. At the

end of the interval, the measurements are passed to the operations support systems.

3. The following measurements may be collected by the SSP:
   * Total number of ABS calls
   * Number of completed ABS calls
   * Number of calling card calls
   * Number of ABS calls that required an operator.

Automatic Call Gapping

Purpose: handles Automatic Call Gapping requests from the SCP.

Processing Functions

1. The SCP sends to the SSP a request for ACG embedded in the message it returns when it detects a certain overload condition due to ABS requests.
2. The SSP initiates ACG as soon as it receives the request.

8.3 Administrative Units in SSP

Trigger Table Administration

Processing Functions

1. Perform initial table load.
2. Provide the trigger indicators and the destination address to which the request for ABS related queries are to be sent.
   The following parameters are required:
   * Trigger indication
   * CCS7 destination address
   * GTT indicator
   * Primary/backup.
3. Perform the following functions:
   CREATE   Make an entry in the table
   MODIFY   Edit fields within a table entry
   CANCEL   Cancel a table entry
   DISPLAY   Display a table entry.
   Note: Table entries are indexed using a key.
   These commands are entered from a local terminal connected to the SSP, or remotely from an operations and support system center.

8.4 Functional Units in SCP

Inquiry SSP/OSS-SCP (Invoke OPDU) and SCP-Response
The processing in the SCP is the same whether the Invoke comes from the SSP or the OSS (Operator Services System). In both cases it will receive an Invoke Operation Protocol Data Unit (OPDU). The SCP sends the response message back to the originator of the query. When the SCP receives an inquiry it checks and analyzes the inquiry request. If the request is formally correct, the SCP performs a read to the LIBD database and the information is sent back with a Return Result message.

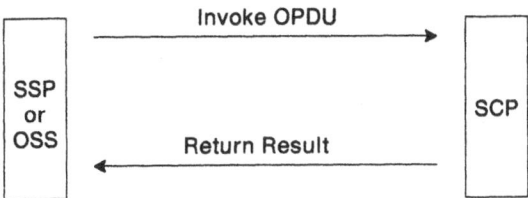

Processing Functions
1. Read the message from application platform input queue.
2. Get the main storage resources.
3. Analyze and process the header/SCCP control data of incoming message.
4. Analyze and process the TCAP control information included in the message.
   The following Invoke types can be used:
   * OLNS (originating line number screening) OPDU
   * Intercept OPDU
   * CC (calling card verification) OPDU
   * BNS (billed number screening) OPDU.
5. Save the header/SCCP control data and TCAP information.
6. Validate the input data of the message.
7. Build a command for record search on LIBD database (key, access parameters, input/output area, return code area, and so on).
   The following keys are supported:
   * For group records = 6 digits NPA-Nxx or RAO-0/1xx,
   * For line records = 10 digits NPA-Nxx-xxxx (called or calling number)
   * For special billing = 10 digits special billing number.
8. Issue database read (the actual read of the database is performed by the application platform).
9. Check and analyze the delivered return code.
10. Record found (no error condition).
    Process the data of the received record:
    a. If the query is for calling card verification (Invoke CC OPDU)

- Record status indicator
- PIN match
- PIN restriction indicator
- PIN service denial
- Service denial indicators
- Calling card subaccount number
- IC-indicators and ICs
- True billing number (optional)

b. If the query is to determine processing of collect or third number call
- Company ID
- Record status indicator
- Collect acceptance indicator (always allowed, always disallowed, or per call approval required)
- Third number acceptance indicator
- Treatment indicator
- Service or equipment indicator
- IC-indicators (U.S. only)
    - Primary preferred
    - Alternate preferred
    - Preferred.

11. Create and format the data part of the output message.
12. Restore the header/SCCP control data, and TCAP information. Determine final header/SCCP control data, and the following TCAP information:
- Package type identifier
- Responding transaction ID
- Component type (Return Result)
- Correlation ID.
13. Build an output message consisting of header/SCCP control data, TCAP control information and data part.
14. Write an output message on the application platform CCS7 output queue.
15. Update counters used for traffic and load (ACG) measurements and statistics.
16. Write the ABS inquiry information on application journal.
17. Free the main storage resources.

Inquiry SSP/OSS-SCP and SCP-Response with Return Error
The SSP/OSS sends a message with the component type Invoke OPDU to the SCP. The request is valid but the SCP is unable to complete the operation, because of improper or invalid data in a component of the inquiry message. The SCP generates a message with the component type Return Error.

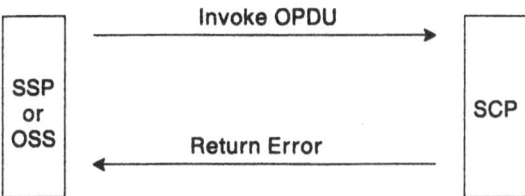

Processing Functions
1. Read the message from application platform input queue.
2. Get the main storage resources.
3. Analyze and process the header/SCCP control data of incoming message.
4. Analyze and process the TCAP control information included in the message.
5. Save the header/SCCP control data and TCAP information.
6. Validate the input data of the message.
7. Invalid or improper data in a component is detected, or the record can not be found in the database.
8. Create an error message, and write the message to the error file.
9. Create and format the data part of the output message.
10. Restore the header/SCCP control data, and TCAP information. Determine final header/SCCP control data, and the following TCAP information.
    * Package type identifier
    * Responding transaction ID
    * Component type (Return Error)
    * Correlation ID
    * Error code
    * Problem data (if any).
11. Build an output message consisting of header/SCCP control data, TCAP control information and data part.
12. Write an output message on the application platform CCS7 output queue.
13. Update counters used for traffic and performance measurements and statistics.
14. Free the main storage resources.

Inquiry SSP/OSS-SCP and SCP-Response because of Protocol Error
The SSP/OSS sends a message with protocol error (probably an application error in the SSP/OSS). The SCP detects the error, writes an error record informing the Telco operator, and sends a message with component type *Reject* to the SSP/OSS.

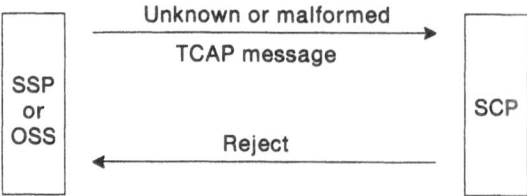

Processing Functions
1. Read the message from application platform input queue.
2. Get the main storage resources.
3. Analyze and process the header/SCCP control data of incoming message.
4. Analyze and process the TCAP control information included in the message.
5. Save the header/SCCP control data and TCAP information.
6. A protocol error is detected (incorrectly formed Invoke OPDU, or component type is unknown).
7. Create an error message, write the message to the error file.
8. Create an operator message and send the message to the responsible ABS operator.
9. Restore the header/SCCP control data, and TCAP information, if possible. Determine final header/SCCP control data, and the following TCAP information.
    *   Package type identifier
    *   Responding transaction ID
    *   Component type (Reject)
    *   Correlation ID
    *   Problem code.
10. Create an output message consisting of header/SCCP control data, TCAP control information, and problem code.
11. Write an output message on the application platform CCS7-output queue.
12. Update counters used for traffic measurements and statistics.
13. Free the main storage resources.

SCP-SSP Automatic Call Gap Embedded in a Return Result Message
An asynchronous CICS transaction is periodically started to measure the load. The parameters for load measurement are the response time of the transactions, the number of traffic counters, and the database load. The ACG can be initiated by an overload situation, or externally by an operator message. The SSP is informed of the overload situation.

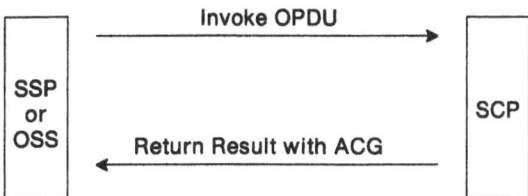

Processing Functions

1. Store the traffic counters from the last measurement period and determine the number of transactions (inquiries from SMS and OSS and updates from SMS).
2. Analyze the database load from this time period.
3. Analyze the application input queue for unprocessed application requests.
4. Analyze the response time counters.
5. Check the number of active application clone tasks for this application.
6. Process these parameters to determine the load. The load and the performance values needed to determine an overload situation are defined in a table. Input values for this table are the above mentioned parameters, or the value of the operator message.
7. If:
   a. It is a normal load situation, no action is required.
   b. There is an overload situation, or ACG initiated by operator, the overload level and sublevel are determined by the table mentioned in 6. The output of processing this table is the overload level and sublevel.
8. Create an output message consisting of header/SCCP control data, and the following TCAP ACG information.
   - Component type (Invoke)
   - Operation code (ACG)
   - Automatic call gap indicators.
9. Write an output message on the application platform CCS7 output queue.
10. Update counters for ACG control and statistics.
11. Write information on ABS application journal.
12. Free the main storage resources.

8.5 Administrative Units in SCP

SMS - SCP Retrieve Application Measurements for ABS

This command retrieves the application measurement for the ABS application, which maintains measurement data. The SMS polls the SCP node for measurement data on a fixed time base. The SCP can generate reports daily, hourly, or every five minutes. The name of the reports, schedules, and so on, are maintained in a parameter control file.

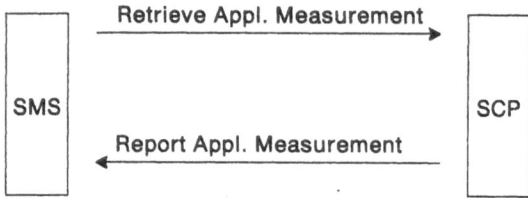

Processing Functions
1. Read the message from application platform input queue.
2. Get the main storage resources.
3. Analyze and process the message data.
4. Save the message.
5. Read the application measurement counters.
6. Using the measurement counters create the following application report message:
    *   Type—application identification
    *   More—indicates that data is sent for one or more intervals
    *   Time interval— beginning and end of the interval
    *   Data regarding the amount of successful responses, response messages with errors, misrouted queries, and queries rejected.
7. Write an output message on application platform SMS output queue.
8. Free the main storage resources.

SMS - SCP Update Processing of LIDB
A update is an addition, change or deletion to the data record. It can be a single transaction or a batch of transactions affecting more than one data record. The update is a logical unit of work; if an abnormal termination or failure occurs, all the changes of a logical unit of work are backed out, either by dynamic transaction backout or emergency restart. The following process assumes that if an SMS update is not acknowledged within a period of time, it will be resent.

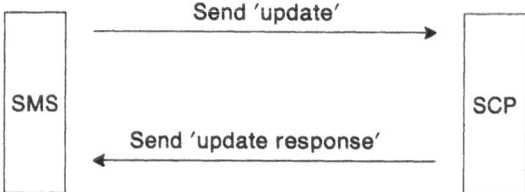

Processing Functions
1. Read the message from application platform input update queue.
2. Get the main storage resources.
3. Analyze, check, and edit the message data.

4. Create a command for a record update on the LIBD.
5. If:
    a. Update was unsuccessful:
        •   Update request was "change" or "delete" an existing record.
            −   Record not found condition.
            −   Sequence number not valid.
            −   Time stamp out of date.
                Write the update request to the error file. Create the response
                "update not successful".
        •   Update request was "add" a new record.
            −   Record exists already.
                Write the update record to the error file. Create the response
                "update not successful".
    b. Update was successful:
        •   The platform data manager saves the previous version of the changed
            record, marks the data file in the process of updating, and sends the
            changed record (or delete or add request) to the update task in the
            subordinate hosts. The update task changes the record and sends the
            "update successful" status. Create the response "update successful".
6. Write a response message to the SMS output queue.
7. Write an ABS update record on the application journal.
8. Free the main storage resources.

ABS Database Customizing and Maintenance
The database is organized as VSAM Key Sequential Data Storage (KSDS). This
gives direct access to records via primary or secondary keys. VSAM KSDS supports
variable length records, reusable space and buffer pooling.

Processing Functions
1. Database description:
    •   Record layout of primary key, secondary keys, and data fields.
    •   Catalog definition into master catalog and user catalog.
    •   VSAM Cluster definition of index and data clusters using:
        −   Control interval size
        −   Control area size
        −   Free space (CI, CA, and so on)
        −   Buffer space
        −   Other parameters.
2. Database initial load
    The load file is sent from the SMS to the SCP via a file transfer program. The
    SCP receives the data and the new records are stored on disk via a VSAM
    utility. This utility creates the index and data areas with the defined CI/CA
    sizes. The data is defined as logical units of work, which means that there are
    possible restart points in the case of an abnormal end.
    The SCP application produces a response file containing the number of
    "loaded" records and a message for each "unsuccessfully loaded" record. The

response file is sent back from the SCP to the SMS via the file transfer program. A similar program is required for mass inserts.
3. Database maintenance and reorganization
Standard batch utilities are used.

- The VSAM catalog statistics display gives information about whether reorganization is needed. The different parameter values giving rise to a reorganization must be defined.
- During reorganization, the database is not available for queries and update requests, and the CICS is shut down. The requests are performed by other SCP components (other CICS systems) with duplicated databases. After reorganization, the CICS is started and a VTAM session with all other SCP components is established. The updates made on another SCP during the reorganization time are incorporated into the reorganized database. All duplicated databases are reorganized using this procedure.

ABS Initialization Routine in the Platform (CICS) Start Up Phase

Processing Functions

1. The start up phase requires an application dependent initialization.
2. The routine must create input, output and save queues, and a Common Work Area (CWA). The CWA is needed for queue pointers, counters and threshold parameter values. The routine sets the counters, threshold and data fields to initial values.
3. The initialization routine loads all application relevant tables, such as the country specific application error message file, and opens all data files to the relevant application.

ABS Application Error Message File

Processing Functions

1. When an application program detects errors, the program must send messages to the operator or to system components. The target of the messages (operator or system component) is defined in a "message filter table" as part of the message file. The text for all application programs and the "filter table" is stored in the error message file.
2. Messages are accessed via a message number. This makes it easy to change the message layout without changing the application programs, and facilitates national language support (for example during the start up phase, when the country specific message file is to be loaded).

Part 4. Emergency Response Service

This part provides detailed information about the Emergency Response Service. It examines what the service is, how it operates in the IN, and the factors to consider when implementing ERS.

This part is ERS-specific; it adds to, and does not replace "Part 1. Overview of the Intelligent Network" of this report.

Chapter 9. ERS Service Description

9.1 Overview

Basic Service

With the **non-IN** Emergency Response Service (ERS), the caller dials a special emergency number (very short, usually three digits) which is uniform within a country, (for example, 911 in the USA, and in West Germany 110/0110 for police and 112/0112 for fire or ambulance).

The central exchange recognizes the special number and the call is connected to a particular emergency response center, which is continuously manned by an operator. A dialog is necessary to determine the nature of the emergency and the appropriate response. The operator may route the call to an emergency station (for example, a fire station) or call the emergency station and request the emergency service needed.

Related basic services include an emergency response information hotline (for poison, suicide, and so on) using country-wide abbreviated dialing.

Enhanced Service

With the **IN** solution, the caller dials the appropriate emergency number as with the basic service. Call origin ANI (Automatic Number Identification) is sent with the call.

In the public switched telephone network (PSTN), the special emergency number triggers the interrogation of a database to determine the real (target) number for call completion.

Translation to the real number depends on such factors as:

* Emergency number dialled
* Originating number (ANI)
* Time and date factors (time of day, day of week)
* Other variables, such as alternate routing in case an emergency station is temporarily closed.

Optional Services

PSTN can send to the emergency response station additional call information such as the ANI, and the time an ERS call was made.

PSTN can access additional databases, such as the "Directory Assistance Service" (DAS) database, to collect additional information.

Emergency response stations can be linked or switched to PSTN, or other public, or private system databases which provide additional information (address of calling party, fastest route to the caller location, where to connect fire hoses, and so on).

Interfaces can be provided to intelligent peripherals (IP) or automatic recording equipment (for example, automatic guiding to store messages for later forwarding or recall).

Background

ERS allows emergency agencies to have access to more complete, accurate, and timely information without relying on the caller, who may be under considerable stress. Service users are keen to enhance home monitoring and increase family safety and protection through ERS. The service establishes a region (country) wide ERS database with advanced information capabilities allowing selective routing and transfer of name and address information to emergency answering organizations.

Benefits

- Service User
  - Enables faster response to emergencies
  - More appropriate response is possible when originating number and other information is passed to the response station, even though the caller may be unable to provide information over the phone.
  - Speed and ease of use; the same number everywhere is easier to remember.
  - Inexpensive; service charge is often included as a local tax on the monthly telephone charge bill of all telephone subscribers.
    For example, $ .16 per month (business), $ .16 per month (residence) in Washington, D.C., USA.
  - The service is normally free from a pay-phone.
- Service subscriber (police, fire, ambulance)
  - Safer response; the service can provide information about dangerous conditions or hazards associated with the proposed action (for example, chemicals are stored at the caller's location).
  - The service can provide routing information to provide the fastest possible route to the destination.
  - Economy and efficiency; automatic routing of calls when emergency stations are closed or busy. Emergency stations answer calls only for their geographic area.
  - Productivity; availability of caller's number reduces abuse of service.
  - Satisfaction of doing a better job.
- Network operator
  - Provide a community service.

   —   Potential allocation of cost for basic IN implementation (CCS7), for the purpose of public welfare and subsequent use of the expertise and network resources for the development and deployment of other IN services.

9.2 Functional Description

General

When a service user makes an ERS call, he is connected to the closest SSP which is either an end exchange or tandem exchange. In the SSP a query to the database (located in the SCP) is triggered. The query is transmitted to the SCP via the CCS7 network and STP nodes.

The ERS application program in the SCP reads a record from the ERS database. The key for the record search is the ANI of the service user (which was included in the query). The ERS application program determines the real number of the emergency center; the Public Safety Answering Point (PSAP). The target number and service logic instructions (such as "send termination") will be sent to the SSP, via the CCS7 network and STP nodes. Communication between the SSP and SCP is done using the TCAP-protocol.

Having received instructions and the target number, the SSP completes the call set up, routing the call through the network to the appropriate emergency response station.

The response may be a "busy" tone because the station is busy, or there is network congestion, or due to station equipment failure. To resolve this, the network should implement:

* SSP retry
* The emergency center forwards the call
* Second query to the SCP to provide alternate station
* Announcement message to the service user.

Service User's Perspective

The ERS user is every person able to operate a service access instrument. Almost every user belongs to a household which has subscribed to PSTN, and in many cases he uses his residential telephone to call the ERS.

The user dials a short number (normally three digits) and is connected to an attendant at the emergency response station enabling verbal communications. ERS provides some functions to the benefit of the user:

* Uniform access code (per country, eventually European or world wide). In some countries there may be different access codes for differing emergency services.
* In many cases the access code corresponds to the number describing the service (for example, in the USA Service 911 has the access code 911).
* The user is connected to the appropriate emergency response station depending on his or her location, day of the week, time of day and other factors.

- Mechanisms provided in case the emergency response station returns a "busy" response mean that the user will always be connected to a responding station.
- The user may use mobile radio phones and coin phones to access ERS.

Service Subscriber's Perspective

Service subscribers are the organizations or agencies such as police-stations, fire-brigades, and ambulance-stations, which must act in case of emergency.

The service subscribers are designated by a municipality to receive and handle emergency calls. In the U.S.A. a receiving emergency agency is commonly called a Public Safety Answering Point (PSAP), also referred to as an Emergency Service Bureau (ESB).

The service subscriber provides a better and faster response if he can enhance the emergency call with information accessed from databases (such as automatic line identification, driving information, hazardous material at the call location). These databases may be owned by institutions like the government, community, or the service subscribers. The availability of ANI reduces abuse of the service considerably.

After ERS call set-up, the attendant at the emergency response station receives ANI information. He uses this information to send a request via a data link or a switched user channel to the ALI retrieval system. It is assumed that the X.25 network is used for communication between the emergency response station and the ALI retrieval system.

The service subscriber may supply some sort of "routing tree" where he specifies to which real number a service user should be connected depending on user origin location, time of day, day of the week and so on. This allows emergency stations to answer only dialed ERS numbers for their geographic area.

In practice the specification of the "routing tree" can be done optionally as follows:

- The service subscriber is connected via a dial-up or leased line to a SMS and uses his data terminal to change the "routing tree" as desired.
- The service subscriber submits a written service request to the Telco, which is then responsible via a service administration center, to update the subscriber's record.
- The service subscriber calls a Telco service center and verbally notifies an operator about the changes desired. The Telco personnel updates the subscriber's record via a service administration center.

The Service Subscriber has the option to assign date and time when changes should become effective in the database.

Using traffic statistics data (for example, number of calls, number of busy situations, number of calls not answered) the Service Subscriber can develop an optimum "routing tree" to prevent the possibility that the user will be connected to an emergency station that is unable to answer his call.

Network Operator's Perspective

When defining IN architecture for ERS use, a network operator must determine whether the Directory Assistance Service (DAS) or an SCP database should be used to provide Automatic Line Identification (ALI).

The DAS alternative is attractive because almost all Telcos have DAS already, and the data stored in the DAS system (directory numbers, names and streets) is exactly the data required for ERS. A Telco planning to implement ERS would have less effort in the initial data acquisition phase and in maintaining their ERS data. The existing DAS must be able to retrieve user records using the telephone number as a key for a record search. There are two proposed ERS architectures for the network operator:

- ERS without access to the DAS database for ALI (Version 1)
- ERS with access to the DAS database for ALI (Version 2).

Version 1 is described in 9.6.2, "Functional Flow" and Version 2 is described in 9.9, "National Dependencies".

The network operator administers the ERS database which is a complex task because data needed for the ERS is stored in other administrative databases, such as the service order database, the directory assistance database, and so on. Updates in any of these databases must be reflected in the ERS database as well. Therefore, procedures must be developed for maintaining all relevant databases. The same must be done for creating the initial version of the ERS database out of available databases.

ERS provides the network operator with scheduled report capabilities to produce statistics reports for himself, and special reports for the service subscribers. These statistics should help the network operator achieve optimum use of his network resources when implementing the ERS.

If the SCP is down, or not available, the ERS call can be routed to a default PSAP. This default routing is performed in the SSP.

Relation to ISDN Supplementary Services

"Calling Line Identification Presentation", one of the ISDN supplementary services, provides the capability needed for ANI, but requires that both ends are ISDN subscribers.

9.3 Standards

Service

Bellcore specifications relating to ERS are described in *E911 Public Safety Answering Point: Interface between a 1/1AESS Centrex Office and Customer Premises Equipment*, TA-TSY-000350.

The E911 architecture described in this document is not IN-based, but the features available with the E911 service are informative. They are:

- Selective routing
- Default routing

- Alternate routing for PSAPs that are traffic busy, on night service, or have a power failure
- Selective, default and alternate routing by the switch, depending on busy conditions, out-of-service, and so on
- Automatic Number Identification (ANI)
- Automatic Location Identification (ALI)/Data Management System (DMS).

Trunk Signaling
A trunk signaling function to pass ANI from a non-SSP exchange to an SSP is required.

9.4 Service Interaction

Service User
The Service User dials the ERS access code, for example, 911. He is connected to the PSAP number without any further interactions.

It should not be necessary in an emergency situation to guide the user in how to set up a call to an emergency response station. However, there is an attendant acting on the call (at the responding station), and he may start a verbal dialog to obtain additional information from the calling person.

Service Subscriber
The ERS database stored in the SCP is used to process an ERS call. Additional databases (such as DAS) owned by emergency agencies or the Telco may be used.

The service subscriber must interact with the SCP to obtain ALI information. When an emergency call is received at the emergency response station, the attendant simply pushes a button to start an ALI request (with embedded ANI), which is sent via an X.25 link to the SCP. The response from the SCP is displayed on the attendant's terminal so he can communicate with the user. Apart from requesting ALI, no further interactions between PSAP and SCP are allowed.

The service subscriber may be authorized to interact with the SMS (via the SMS interface) for updating, reporting, and query activities on the ERS database.

However, the service subscriber will probably not be allowed to directly interact with any other Telco-owned database for security reasons. If the ERS subscriber is so authorized, he uses a Telco specified interface which is not part of the IN architecture.

When interacting with databases from emergency agencies, it is the subscriber's responsibility to negotiate a proper interface. This may be any commercially available implementation.

There should be no dialog between the service subscriber and the SCP, and the request for ALI information should be kept simple. To setup a request for ALI, the ANI information should be taken from the emergency call received automatically by the PSAP terminal.

If a subscriber is authorized by the Telco to access the SMS he will be guided by an interface. This is described in 2.5.2, "Service Subscriber". No data is required by the service subscriber in his interactions with SCP.

Interacting with the SMS, the service subscriber may enter the following data:

- Service subscriber name
- Address
- Security passwords
- Routing parameters (for example, time of day, day of week)
- PSAP phone numbers to which calls should be routed
- Dates to schedule update events
- Specifications for report generation
- Parameters to sample subscriber related service data.

Network Operator

The network operator must maintain the user's data stored in the SCP or SMS (or in any other Telco database for example, DAS).

To access DAS or service order databases, a Telco uses existing interfaces that are not relevant to the ERS architecture.

The data which the network operator must provide is the same as that for the service subscriber. In addition, a network operator must supply information on all service user related data, such as:

- Telephone number (ANI)
- User's name
- Full address
- Area code, community, county, and state
- Public Safety Answering Point (PSAP) number
- Class of user (coin, residence, business).

Access Instrument

The ERS can be accessed with any instrument that can be connected to the PSTN. This includes residential and business telephones, coin phones, PBX attached phones, mobile radio phones, and special emergency alarm instruments.

The service subscriber must have the CLASS-feature "Display of ANI", or ISDN access, and the appropriate terminal.

To obtain additional information, special terminals can use ANI of the calling user for the database access, and provide authorization for the database access (for example, using the service subscriber's ANI).

9.5 Billing

Billing requirements will vary substantially from country to country.

It can be assumed that an ERS call is free for service users and no billing is done on a per call basis. Usually, the public pays for the ERS service, for example, the telephone subscriber will be charged a monthly amount. To calculate this amount (which must cover a Telco's cost in providing ERS), ERS traffic measurements statistics can be used.

SSP
No specific ERS billing is required.

SCP
No specific ERS billing is required.

SMS
Because of the social aspect of the ERS, a Telco will normally have a special contract with the emergency service-providing institutions to cover the eventual billing of activities performed by a subscriber in the SMS.

9.6 Service Logic

9.6.1 Distribution

SSP
The SSP database contains triggers so it can detect that a call needs a special routing service. The SSP must contain logic to detect these triggers and to send an inquiry message to the SCP containing the ANI. When the SCP responds, the call set up is completed by the SSP.

Service logic is also required to handle other types of messages from the SCP, or network control information (such as Automatic Call Gapping), which can be included in a TCAP message response.

The IN service logic must be able to coexist with traditional ERS logic.

SCP
The SCP must contain service logic so it can determine the real destination number from an incoming originating number, depending on the individual service subscriber routing parameters. This service logic must interpret the TCAP message (with imbedded ANI) and access an ERS database record, using the call originating number (ANI) as a key.

Having retrieved all necessary data, the SCP creates the output message containing the target (PSAP) number and ANI information. The message is enhanced with TCAP protocol data units, and sent via the CCS7 network to the SSP.

If the SCP receives a request from an PSAP attendant to provide Automatic Line Identification (ALI), it interprets the incoming message, which was received via an X.25 link. The SCP then reads the ERS database using the delivered ANI as the key for a record search, and creates the output message containing the ALI information. This message is then sent to the PSAP via the X.25 network.

The SCP must also contain service logic to handle database updates, collect traffic data and exchange transactions with the SMS, as well as to handle exceptional conditions, such as overload situations, database errors, and no number assigned (for example, key not found).

Triggers

The triggering data for ERS is the access code or the dedicated line (for special alarm systems connected without dialing), and the query is started as soon as the complete access code is received.

The trigger table in the SSP is administered by the Telco and describes which part of the data will trigger a query to the SCP, and at which point in the call the query is started.

The triggering, routing, and contents of the query can be based on any parameters available to the SSP at the time the query is sent.

9.6.2 Functional Flow

SSP - SCP

Version 1: ERS with access to the SCP database for ALI

Figure 33 shows the flow of messages when setting up an ERS call. It consists of the following steps:

1. The user dials the ERS access code (for example 911) which triggers a query in the SSP to the SCP. The relating TCAP Invoke message contains the caller's number (ANI).
2. Based on the delivered ANI, the SCP accesses its ERS database to get the corresponding user record. The selective routing logic of the application program determines the telephone number of the PSAP to which the user should be connected.
3. The SCP communications program creates a TCAP Return Result message containing the ANI, and the PSAP telephone number. This is sent to the SSP.
4. The SSP completes the call, routing it to the proper PSAP attendant for further handling of the emergency call. The attendant receiving the call is provided with the caller's phone number, with which he starts a data inquiry to retrieve Automatic Location Identification (ALI).
5. The ALI request (with embedded ANI) is sent via an X.25 link to the SCP.
6. The SCP will access its ERS database to get the user record corresponding to the ANI. Name and address information in the user record is used for ALI.
7. The SCP communications program creates the responding output message containing ANI and ALI information. This is sent back to the PSAP. The attendant receiving the call is provided with the caller's phone number, his

name (if the caller uses his own telephone) and location (normally the location of the emergency situation). This information is displayed on the attendant's screen.

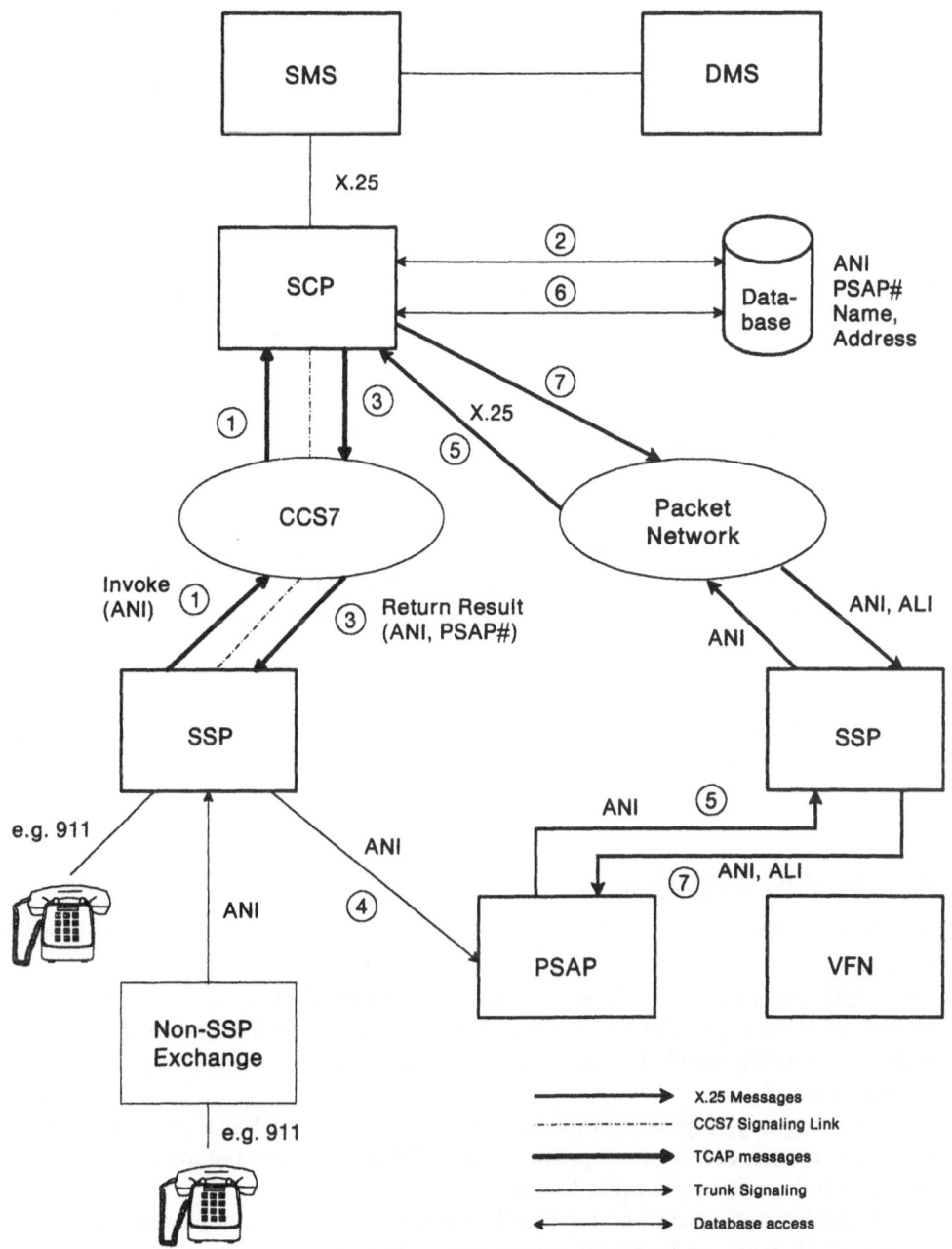

Figure 33. ERS Version 1, with Access to the SCP Database for ALI

SCP - SMS

No specifications for ERS describing SCP - SMS messages were available.
However, there will be transactions between these network elements to provide for:
- ERS database updates
- Reporting
- Statistics
- Status information.

9.7 Traffic Measurement Requirements

There are no known specific ERS requirements. Generic traffic measurement
requirements are described in Chapter 2, "Functional Characteristics Common to
Selected IN Services".

9.8 Dynamic Requirements and Performance

SSP

The ERS user program should begin timing for a response from the SCP after
sending a query message. A nominal value of three seconds should be used. If a
response from the SCP is not received in this period, the service user should be
connected to a default PSAP.

SCP

On receipt of a query from the SSP, the SCP should respond within one second on
average, and within 1.5 seconds 99% of the time.

9.9 National Dependencies

General

The establishment of a Public Safety Answering Point (PSAP), or an emergency
response station is governed by legal, geographic, and social factors, so that the
number of PSAPs established in a country will depend on the country specific
environment.

An emergency response station may be linked on-line to PSTN, or other public,
or private system databases to provide additional information (address of calling
party, fastest route to the caller location, and so on).

Subscriber authorization to access Telco-owned databases, not stored in the
SCP/SMS, depends on legal restrictions due to data security.

The Data Management System (DMS) required to build and update the ERS database depends on Telco administrative databases such as DAS, service order data, street address guides.

The billing requirements for ERS are very country specific. A Telco will have special contracts with the emergency service providing bureaus to bill activities performed by a subscriber in the SMS.

Use of mobile radio phones and special emergency alarm instruments may be used in some countries.

Each country must provide mechanisms to avoid ERS abuse calls.

Using the DAS database for retrieval of ALI (Version 2)

Figure 34 shows the flow of messages when setting up an ERS call. It is assumed that an X.25 network is used to set up communication between PSAP and DAS. The functional flow consists of the following steps:

1. The user dials the ERS access code which triggers a query in the SSP to the SCP. The relating TCAP Invoke message contains caller's number (ANI).
2. Based on the ANI, the SCP accesses its ERS database to get the corresponding user record. The selective routing logic determines the telephone number of the PSAP to which the user should be connected.
3. The SCP communications program creates the responding TCAP Return Result message containing ANI, and the PSAP telephone number. This is sent back to the SSP.
4. The SSP completes the call, routing it to the proper PSAP attendant for further handling of the emergency call. The attendant receiving the call is provided with the caller's phone number, which he uses to retrieve ALI from the DAS.
5. The ALI request (with embedded ANI) is sent via an X.25 link to the DAS system.
6. Based on the ANI, the DAS system accesses its database to get the corresponding user record. Name and address information in the user record is used for Automatic Location Identification (ALI).
7. The DAS communications program creates the output message containing ANI and ALI information. This is returned to the PSAP. The attendant receiving the call is provided with the caller's phone number, his name, and location. This information is displayed on the attendant's terminal.

The ERS record must contain the telephone number of the calling user, and the real (PSAP) numbers and routing parameters.

The caller's number provided in the query to the SCP (ANI) is used as a key to find the PSAP number to which the call must be transferred (selective routing).

Using the SCP to Access the DAS Database:

In some countries there may be legal, or data-security restrictions preventing the service subscriber from accessing the DAS database directly. In such cases, it may be desirable that the emergency response station can send an ALI request via the X.25 network to the SCP. To answer the request, the SCP sends an inquiry to the DAS system which in turn provides ALI information. The following process would take place:

1. The user dials the ERS access code which triggers an SSP query to the SCP. The TCAP message contains the number of the caller (ANI).

2. Based on the ANI, the SCP accesses its ERS database to get the corresponding user record. The selective routing logic determines the telephone number of the PSAP to which the user should be connected.

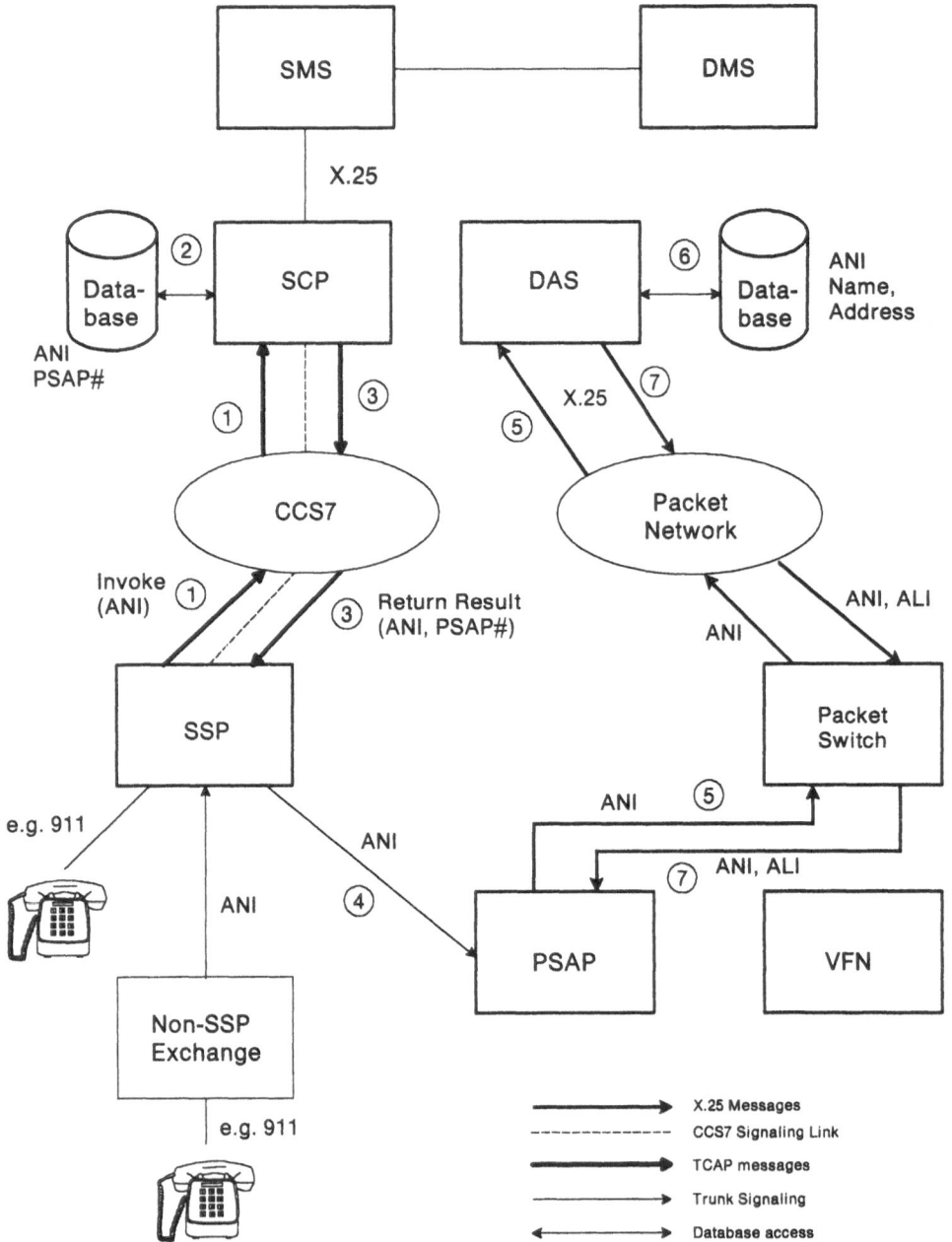

Figure 34. ERS Version 2, with Access to the DAS Database for ALI

3. The SCP communications program creates the TCAP Return Result message containing ANI, and the PSAP telephone number. This is sent back to the SSP.
4. The SSP completes the call, routing it to the proper PSAP attendant. The attendant receiving the call is provided with the caller's number, which is used to start a data inquiry to retrieve ALI via the SCP from the DAS database.
5. The ALI request (with embedded ANI) is sent via an X.25 link to the SCP.
6. The SCP communications program must save the ANI and PSAP number, and send a message via an X.25 link to the DAS system. This message contains the ANI parameter information.
7. Using the DAS system, an application program accesses the DAS database to the user record defined by the ANI. The caller's name and address are taken for ALI.
8. The DAS communications program sends a response message to the SCP, including the ALI data.
9. The SCP communications program combines the ALI message with the previously saved ANI and PSAP number, and creates the responding output message. This is sent back to the SSP.
10. The SCP sends the response message via an X.25 network to the PSAP attendant. The attendant receiving the message is provided with the caller's number, name, and location. This information is displayed on the attendant's terminal.

ERS Version Selection Criteria
The selection of one of the described ERS versions depends on individual country requirements. A Telco must consider:
*   The efforts required to build, load, and update the ERS database
*   Procedures to ensure data integrity in more than one database
*   Disk storage capacity to store necessary data
*   The number of components required to handle a call for each version of ERS
*   Impact on the DAS system (such as availability and performance) if used for an additional service
*   Response time to set up an emergency call
*   DAS capability to retrieve user records using the telephone number as key for a record search
*   Compatibility with or relationship to existing emergency call services
*   Compatibility for ERS enhanced features (such as storage of personal data).

9.10 Future Considerations

The Telco can store emergency contact information such as the user's blood type and allergic reaction data, accessible to an emergency response station. The information should also be accessible to a PSAP operator when the call comes from a location other than the caller's home.

Access to ERS via a mobile radio telephone may be valuable. An enhanced version of the ERS could be based on a new network capability to transport Automatic Location Identification (ALI) from the SCP via the SSP to the emergency response station. This capability must be implemented in all exchanges between the SSP and the emergency response station. It would be based on the insertion of data in CCS7 messages, which requires definition of corresponding standards because of the multi-vendor environment in most countries.

This enhanced version would require the same ERS database in the SCP as described for version 1. The main advantage would be a reduction in the number of transactions to complete an ERS call, a faster call set up, and complete delivery of all information required at the attendant's station as soon as he responds to the call.

Version 3

ALI information is retrieved from the SCP and provided at ERS call set-up. As shown in Figure 35, there are only four steps in setting up an ERS call:

1. The user dials the ERS access code which triggers an SSP query to the SCP. The relating TCAP Invoke message contains the number of the caller (ANI).
2. Based on the ANI, the SCP accesses its ERS database to get the corresponding user record. The application program selective routing logic determines the telephone number of the PSAP to which the user should be connected. The user's name and address is used for ALI.
3. The SCP communications program creates the responding TCAP Return Result message which will contain ANI, ALI, and the PSAP telephone number, which will be sent to the SSP.
4. The SSP completes the call and routes it to the proper PSAP attendant. The attendant receiving the call is provided with the caller's number, name, and the location of the caller. This information is displayed on the terminal.

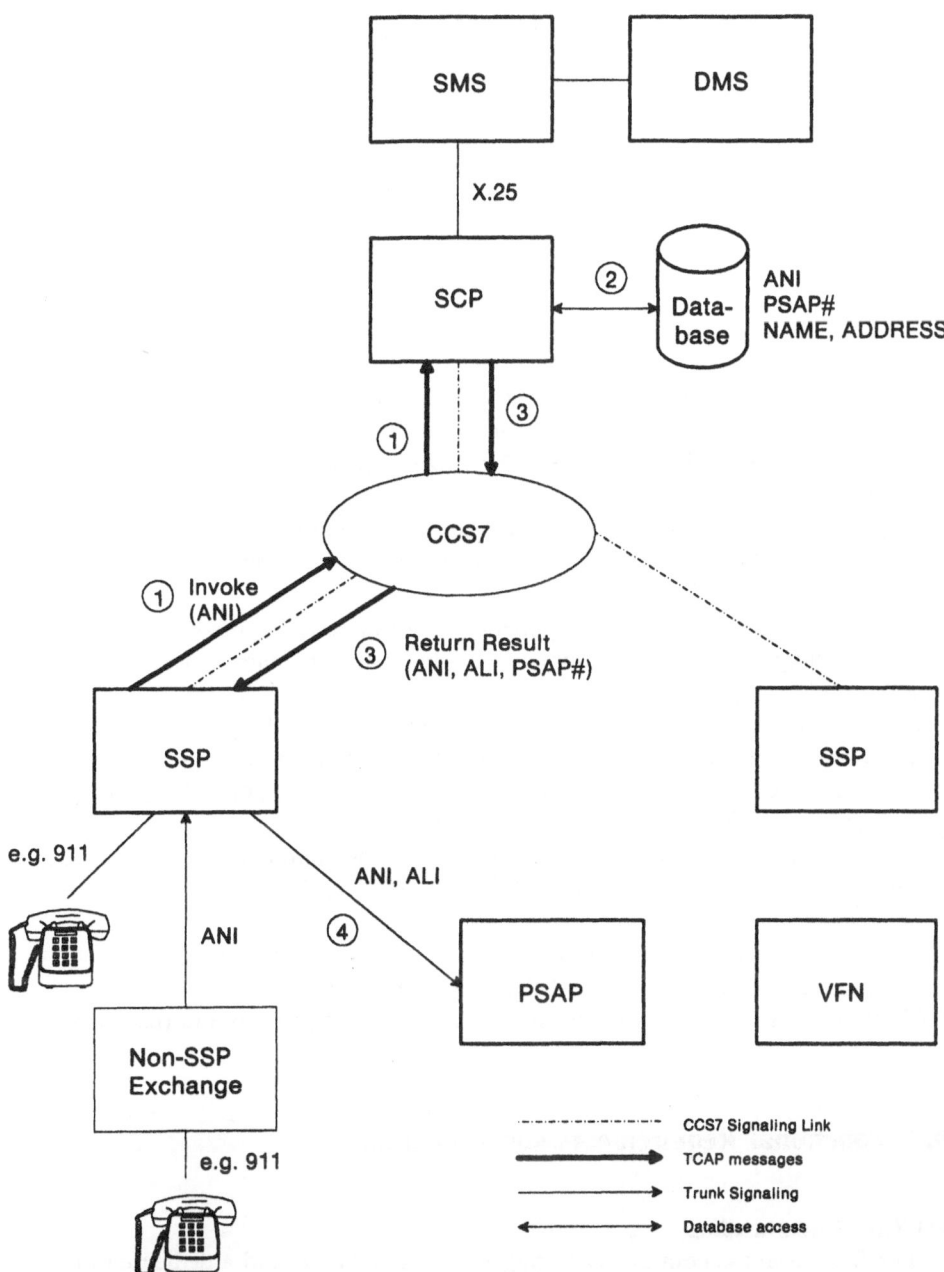

Figure 35. ERS Version 3, ALI Retrieved from SCP at Call Set-UP

Chapter 10. ERS Application Description

The description in this chapter is based on IBM's and Siemens' views of:

- Functional requirements and allocation. Requirements are specified and each item is assigned to the SCP or the SSP depending on where it must be developed.
- Functional units. Processing functions needed to handle the requested tasks are described and grouped logically into functional units for the separate components.
- Administrative Units. Processing functions not directly required in the real time call handling process are combined into administrative units, separately for the SSP and SCP.

A bold printed SSP means that the associated function must be performed in the Service Switching Point. A bold printed SCP means that the associated function must be performed in the Service Control Point. In the Emergency Response Service application description, it is assumed that:

- TCAP is the application layer protocol used in the communication between the Public Safety Answering Point (PSAP) and the SCP via an X.25 network (PSPDN).
- From an application point of view it is not relevant whether a TCAP transaction is transmitted via a CCS or X.25 network.

TCAP messages in the SCP are described in a similar way to those in the GNS.

10.1 Functional Requirements and Allocation

Real Time Call Handling

1. The SSP database contains two triggers; an access code and a dedicated line, enabling it to detect that a call needs a special routing service. The SSP triggers a query to the database located in the SCP. **SSP**
2. The TCAP invoke message contains the caller's number and the query is transmitted to the SCP via the CCS7 network and STP nodes. **SSP**
3. The SCP contains service logic to interpret an incoming TCAP message, and extract the caller's number. **SCP**

4. The ERS application program in the SCP reads a record from the ERS database. The key for the record search is the service user's number included in the query.   **SCP**
5. The ERS application's selective routing program determines the target number (that is, the number called). This may depend on parameters, such as time, day, and so on.   **SCP**
6. Having determined all necessary data, the SCP must create the output message containing the target and calling number.   **SCP**
7. The output message must be extended with TCAP protocol data (return result) units and sent to the SSP via the CCS7 network.   **SCP**
8. Having received instructions and the target number, the SSP completes the call, routing it through the network to the appropriate emergency response station.   **SSP**
9. Solutions should be provided to resolve a "busy" status, for example if a station is busy, the network congested, or if station equipment has failed;   **SSP**
    - SSP retry
    - Call forwarding
    - Second query to SCP to provide alternate station
    - Announcement message to the service user.
10. In the emergency response station an Automatic Location Identification (ALI) request is sent to the SCP via an X.25 link.
    The application interprets the incoming message and, based on the ANI request, the SCP accesses its ERS database to get the corresponding user record. Name and address data in the user record provides automatic location identification.   **SCP**
11. The SCP communications program creates a responding output message containing calling number and ALI information. The message is sent back to the PSAP (via an X.25 link).   **SCP**
12. The SSP must provide the following service features: **SSP**
    - Default routing
    - Alternate routing for PSAPs that are busy, unattended, or that have a power failure
    - Selective, default and alternate routing to be performed by the switch
    - Automatic number identification (ANI).
13. The SCP must provide Automatic Location Identification.   **SCP**
14. These are TCAP messages from the **SSP to SCP:**

INVOKE	Initial query from SSP (provide instruction)
INVOKE	Report error
REJECT	Protocol error detected
RETURN RESULT	Termination information (for example, data for statistics).

15. These are the TCAP messages from the **SCP to SSP:**

RETURN RESULT	Response containing routing information (In U.S. this is done with invoke (connection control))
RETURN ERROR	Error in data detected

| REJECT | Protocol error detected |
| INVOKE | Automatic call gapping. |

SMS-SCP Transactions

1. Transactions between SCP and SMS must provide updating, reporting, statistics, and status information.   **SCP**
2. The SCP must contain service logic to handle database updates.   **SCP**
3. The network operator must be able to produce ERS statistic reports for himself, and special ERS reports for service subscribers.   **SCP**

Traffic Measurements

1. It is anticipated that ERS specific counts will take place in the SSP. These counts include: **SSP**
   - Total number of calls
   - Number of completed calls
   - Number of resolved "busy" call situations
   - Number of unresolved "busy" call situations.
2. It is anticipated that ERS specific counts will take place in the SCP. These counts include: **SCP**
   - Total number of queries
   - Number of queries by emergency response station.
3. A number of peg and usage counts should be provided. Administration reports should be recorded every 30 minutes.   **SSP**
4. Application measurements are made in the same 30-minute intervals as the node measurements are collected, beginning on the hour. At the end of the interval, the application passes the measurements to the operations and support system.   **SSP**

Exceptions

1. Logic must be included to process exceptional conditions, such as overload situations and database errors.   **SCP**
2. The SCP node detects an overload by measuring the delays between receiving a query and returning a response to the query. When the SCP node determines this overload, it notifies the application with a message. The application should respond with certain actions to help alleviate the overload. The action taken depends on the level of overload which is indicated to the application by the SCP (see 2.11.3, "Overload Handling").   **SCP**
3. Service logic must also exist to handle other messages from the SCP, or network control information (such as automatic call gapping), which is based on the destination and included in a TCAP message response.   **SSP**

Supporting Functions
1. The ERS application must support the initial loading process of the ERS database.   **SCP**
2. The SMS produces load files when loading the SCP databases. These files can be transferred to the SCP via an X.25 link or by tape transport, to create the initial service database. The application must generate a response file, containing one response message for each message in the SMS load file. The response file is written onto magnetic tape and transferred to the SMS (X.25 link or tape transport).   **SCP**

10.2 Functional Units in SSP

Initial Call Handling
This unit is responsible for initial ERS call handling.

Processing Functions
1. To access an emergency center, the caller dials a special access number or originates from a specially dedicated line.
2. The SSP switch function program (call processing) in the LTG collects the digits, and sends the digit block to the digit analysis program in the CP.
3. The digit analysis program in the CP (based on the trigger table) identifies that this number requires special treatment (that is, it requires further translation by an SCP).
4. The CP assembles all the destination related information from the database, including the GTT indicator, and DPC status.
5. The Coordination Processor sends a TCAP request to the ERS Service Unit in the LTG, providing the DPC of the SCP to which the query will be sent, or the DPC of the STP that will perform the global title translation.

ERS Service Unit
This unit handles all ERS requests to an SCP database and all responses from the SCP. This unit normally resides in the LTG.

Processing Functions
1. Verifies the DPC status.
2. Assigns the transaction ID and invoke ID for the Provide Instruction operation and formulates the transaction component Invoke and passes it to TCAP.
3. The Invoke (Provide Instructions) contains:
    *   Calling number or origin
    *   Number called
    *   Exchange ID
    *   LATA ID.
4. The message is sent using the platform's TCAP sending capability, which uses the CCS7 lower layers (SCCP class 0, and MTP) to deliver the message to the destination SCP. TCAP formulates the query, provides component and

transaction level processing, and passes it to the SCCP. TCAP also associates the transaction with the appropriate reply at the transaction level and the correlation at the component level.

5. The SSP waits three seconds for a response before timing out.

6. In case of timeout, the SSP may send an announcement or reorder tone to the caller.

7. The response message from the SCP passes through the MTP, SCCP and TCAP before arriving at the application. MTP and SCCP provide transaction and deliver the message to the application. Three basic response types are expected from the SCP, namely:

Invoke (connect call)	Used to return a normal result (U.S. only)
or	
Return Result	Used to return a normal result (CCITT)
Return Error	Used to indicate an error in handling the query (normally due to an SCP error)
Reject	Used to reject a query due to message or format error

8. On receiving a "Return Error" or "Reject" response, the SSP handles the call as if it were the timeout case.

9. On receiving a normal response (return result or "connect call" invoke), the SCP returns any of the following:

    *   A translated number, or a call treatment indicator if unsuccessful in translating the number or the call is not allowed.
    *   Automatic call gapping
    *   Optional request for call termination information.

10. The SSP uses the translated number to complete the call as a normal call. This is usually done by reinitiating the translation procedure with the new translated number.

11. For emergency calls, the SSP will **not** produce an AMA record for each completed call. Billing for emergency calls is normally done on a fixed monthly charge, agreed upon with the local authorities.

12. The appropriate traffic counts are pegged by the standard call processing routines.

13. If the SCP returns an ACG response, the appropriate call gapping parameters are sent to the unit that handles call gapping. This normally results in the SSP stopping any pending requests to the SCP for the duration of the gapping interval.

14. The SSP sends a return result message containing call termination information to the SCP at the end of the call (if the SCP requested it).

15. If the SSP was unable to complete the call as specified in the reply message from the SCP, it sends the following responses to the SCP:

Invoke (Return Error)	If the SCP did not request any call termination information.
Return Error	If the SCP requested call termination information.

16. If the SSP detects a protocol error in the response message from the SCP, it sends a "Reject" response message to the SCP. This assumes that the SSP has found enough information to include in the reject message.

Busy Handling
This function handles inability to complete a call to the designated PSAP.

Processing Functions
1. The SSP may be unable to complete the emergency call to the PSAP center designated in the SCP reply message because the station is busy, the network is congested, or because there is station equipment failure.
2. The SSP attempts to resolve such situations by implementing:
   - SSP retry
   - Call forwarding to a default PSAP
   - Second query to the SCP to provide an alternate station
   - Announcement message to the service user.
3. The SSP should include default routing and alternate routing for PSAPs.

Traffic Measurements
This function maintains all required traffic and maintenance measurements.

Processing Functions
1. Peg and usage counts are provided with a recording period of 30 minutes.
2. Application measurements are made in 30-minute intervals on the hour. At the end of the interval, the measurements are passed to the operations support systems.
3. The following measurements may be collected by the SSP:
   - Total number of ERS calls
   - Number of completed ERS calls
   - Number of resolved "busy" ERS calls
   - Number of unresolved "busy" calls.

Automatic Call Gapping
This function handles Automatic Call Gapping requests from the SCP.

Processing Functions
1. The SCP sends the SSP a request for Automatic Call Gapping embedded in the message it returns when it detects a certain overload condition to a particular ERS destination.
2. The SSP initiates ACG as soon as it receives the ACG request.

10.3 Administrative Units in SSP

Trigger Table Administration

Processing Functions
1. Perform initial table load.
2. Provides the trigger indicators and the destination address to which the request for ERS related queries are to be sent.

     The following parameters are required:
    *   Trigger indication
    *   CCS7 destination address
    *   GTT indicator
    *   Primary/backup.
3. Performs the following functions
     CREATE   Make an entry in the table
     MODIFY   Edit fields within a table entry
     CANCEL   Cancel a table entry
     DISPLAY   Display a table entry.
     Note: Table entries are indexed using a key.
     These commands are entered either from a local terminal connected to the SSP, or remotely from an operations support system center.

10.4 Functional Units in SCP

There are three types of functional units:
1. TCAP Input Message Handling
2. TCAP Message Processing
3. TCAP Output Message Creating.
Because the functions involved in TCAP input and output message handling are very similar for ERS TCAP messages, only one functional unit is described for "TCAP Input Message Handling" and "TCAP Output Message Creation".

     The relationship between the functional units and their interface to the application platform is shown in Figure 36.
All functional units use common application platform interfaces, access the database in much the same way, and work on closely related input and output data. Their common capabilities are:
1. **Interfaces to Application Platform and Queues**
    *   CICS Temporary Storage Read Next command to read input message from the ERS application's inquiry queue.
    *   CICS Temporary Storage Write command to write output message on the ERS application's CCS7 output queue.
    *   CICS EXEC LINK facility to access an ERS database record, and write the ERS inquiry application journal, or the ERS error file.

2.   **Database Access**
  * One record is read via keyed access from the ERS database, which is a VSAM KSDS data file. The key for a record search is built using the caller's information, which is supplied in the data element of the incoming TCAP inquiry message.
  * To improve performance, parts of the file or the whole file can be kept in main storage. The application platform controls the data storage method.
  * Records are written on the ERS journal file or the ERS error file, which is a VSAM ESDS data set.

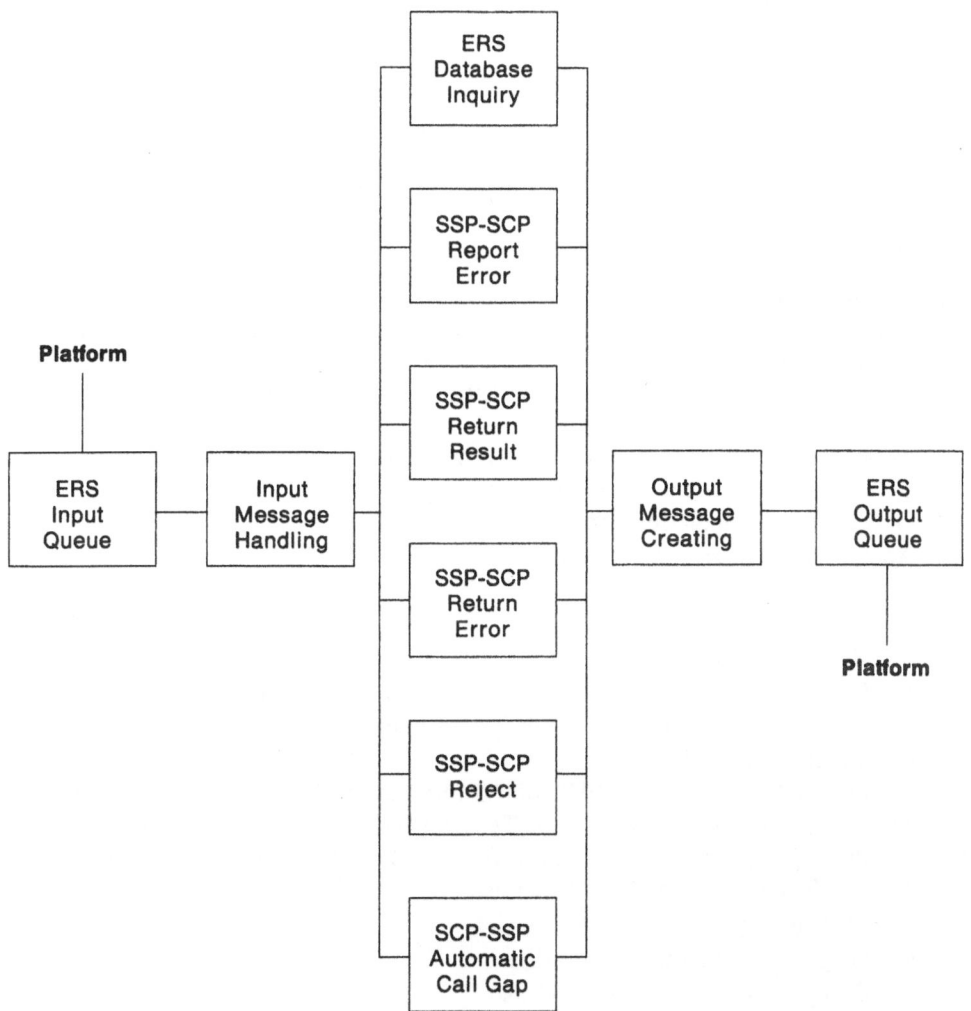

Figure 36. SCP Functional Units and Application Platform Interfaces

3. **Input and Output Data**
   - The TCAP part of the incoming message contains coded information about the actions required by the ERS application program.
   - The data element of the input message consists of information about the originating call area.
   - The data element of the output message includes information about the originating call, and the determined PSAP number, or, in case the input message was received from an emergency response station, the appropriate ALI data.
   - The TCAP part of the output message supplies coded information for the functions required by the SSP, or the PSAP.

TCAP Input Message Handling

A TCAP input message is received by the SCP, processed by the application platform and written on an ERS application input queue. The handling of an application TCAP input message starts with reading the message from the queue. The message is then interpreted and analyzed. If there are no TCAP protocol errors, the data element of the message is given to the next application unit, depending on the message type.

Processing Functions

1. Read message from ERS application input queue. A CICS Temporary Storage Read Next command must be issued to receive the message. The queue accessed by this command is allocated to the ERS application and used only for incoming TCAP messages.
   Note: There might be two input queues; one for input messages from the CCS-network, and one for messages coming into the SCP via the X.25 network.
2. Get the main storage resources.
3. Analyze and process the header/SCCP control data of incoming message. Data analyzed in this process is:
   - Message type
   - Routing label
   - Called party address (ERS application sub system number)
   - Calling party address (SSP sub system number)
   - TCAP message length.
4. Analyze and process the TCAP control information included in the message. The following is a summary of the TCAP message contents to be analyzed (a detailed description can be found in Bellcore TR-TSY-000064 FSD 31-01-0000, TCAP Message Format for 800 Service):
   - Package type identifier
   - Originating transaction identifier
   - Component types and identifiers
   - Operation codes
   - Error codes
   - Problem codes
   - Parameter set (for example, ANI).

5. Save the header/SCCP control data and TCAP control information.
6. If a TCAP protocol error was found in the analyzing process:
   - Create an error message and a command to write the error message on an ERS error file. Include the information in the write command:
     - ERS error file name
     - Address of the field containing the error message
     - Access parameters (for example, type of access)
     - Return code area.
   - Issue a write command to the ERS error file, which is a VSAM ESDS file. The CICS Exec Link facility is used to access the error file. The application platform controls and monitors the error file.
   - Having written to the error file, check and analyze the delivered return code.
   - Create a warning message to be sent to the Telco ERS operator.
   - Write the warning message on an application output queue. A CICS Temporary Storage Write command must be issued to put the message on the queue. (There should be a particular ERS application output queue to receive all Telco ERS operator messages).
   - Create and format the data element of the TCAP output message, which relates to the protocol error.
   - Call the TCAP Output Message Creating function.
7. If the analyzing process has found no errors, call the appropriate processing unit and provide TCAP application input data to this unit.

TCAP Output Message Creating
This part of the ERS application starts when a processing function has finished, and requires that a TCAP message be sent to an SSP. The TCAP output message is created and written on the ERS application output queue for subsequent processing by the application platform data communications part.

Processing Functions
1. Receive output data supplied by the processing unit.
2. Restore TCAP control information. Determine and format the TCAP output message. The following summarizes the contents of output message: (Details can be found in Bellcore TR-TSY-000064, FSD 31-01-0000, TCAP Message Format for 800 Service).
   - Package type identifier
   - Responding transaction identifiers
   - Component types and identifiers
   - Operation codes
   - Error codes
   - Problem codes
   - Parameter set, for example ANI, PSAP Number, ALI.
3. Restore the header/SCCP control data and determine final header/SCCP control data, containing the following information:
   - Message type

- Routing label
- Called party address
- Calling party address
- TCAP message length.

4. Create an output message consisting of header/SCCP control data and the TCAP message.
5. Write an output message in the ERS application output queue. A CICS Temporary Storage Write command must be issued to put the message in the queue. A queue is allocated to the ERS application and only used for outgoing TCAP messages.
   Note: There might be two queues for TCAP output messages, one for CCS7 output and one for output to the X.25 network.
6. Update counters used for traffic measurements and statistics.
7. Write relevant information, collected during the ERS logic process, on the ERS application journal. The journal is a VSAM ESDS data file and is written using the CICS Exec Link facility. The application platform controls and monitors the ERS journal.
8. Free the main storage resources.

SSP-SCP ERS Database Inquiry

A TCAP message Invoke (Provide Instructions) is received. This means that:

- The inquiry data is valid and the query to the ERS database is successful. The ERS application program delivers parameters and information for the generation of a TCAP output message with the component type Invoke (Connection Control) in the subsequent TCAP output message creation unit.
- The message data is not valid, or the query to the ERS data base is not successful. The ERS application program delivers data and information to generate a TCAP Return Error output message in the subsequent TCAP output message creation unit.

Note: The CCITT standards define Return Result instead of Invoke (Connection Control). In the following descriptions, these TCAP component types are synonymous.

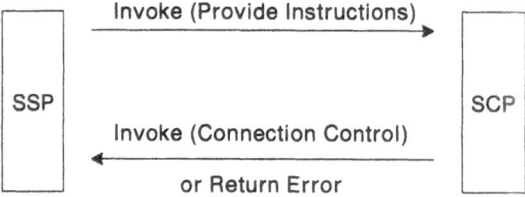

Processing Functions

1. Check and validate ERS relevant input data of the received TCAP message.
2. If validation is unsuccessful, provide component type and error code. Call TCAP Output Message Creating function.
3. Create a command for a record search on ERS database. The command must include:
   - File name
   - ANI as key for record search, which is supplied by the input message
   - Input/output
   - Access parameters
   - Return code area.
4. Issue a read command to the ERS database, which is a VSAM KSDS data file. The CICS Exec Link facility is used to access the ERS data base. To improve performance, parts of the ERS file can be kept in main storage. The application platform data manager controls the storage method.
5. After completion of the ERS database read, check and analyze the delivered return code.
   If record found (no error condition):
   - Process the data of the received record to determine the PSAP number, considering such routing parameters as day of week, time of day, and so on. Determine the component type of the output message to be sent.
   If record not found (error condition):
   - Determine the component type and error code (such as data unavailable).
   - Create an error message and a command to write the error message to an ERS error file. Include the following information in the write command:
     − ERS error file name
     − Address of the field containing the error message
     − Access parameters
     − Return code area.
   - Issue a write command to the ERS error file, which is a VSAM ESDS file. The CICS Exec Link facility is used to access the error file. The application platform controls and monitors the error file.
   - After completion of the error file write, check and analyze the delivered return code.
6. Create and format the data element of the TCAP output message, which is specific for the ERS database inquiry.
7. Call the TCAP Output Message Creating function.

SSP-SCP Invoke (Report Error)

The SSP could not complete an ERS call with the message received from the SCP after an initial inquiry. The return message from the SCP defined "No Sending Termination Information". The SSP sends a TCAP *Invoke (Report Error)* message to the SCP.

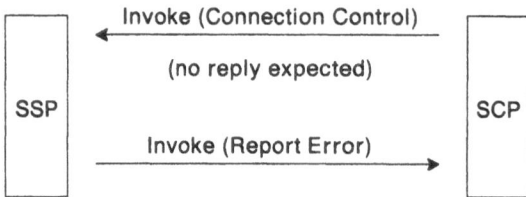

Processing Functions
1. Process the relevant error data from the TCAP input message.
2. Create an error message and a command to write the error message on an ERS error file. The command must include the following information:
    *   ERS error file name
    *   Address of the field containing the error message
    *   Access parameters
    *   Return code area.
3. Issue a write command to the ERS error file, which is a VSAM ESDS file. The CICS Exec Link facility is used to access the error file. The application platform data manager controls and monitors the error file.
4. After completion of the error file write, check and analyze the delivered return code.
5. Depending on the TCAP error and return codes, create a warning message to be sent to the Telco ERS operator.
6. Write the warning message on an application output queue. A CICS Temporary Storage Write command has to be issued to put the message on the queue. There may be a particular ERS application output queue to receive all Telco ERS operator messages.
7. Update the counters used for traffic measurements and statistics.
8. Free the main storage resources.

SSP-SCP Return Result
The SSP could complete an ERS call with the message received from the SCP after an initial inquiry. The return message from the SCP defined "Sending Termination Information". The SSP sends a TCAP Return Result message to the SCP.

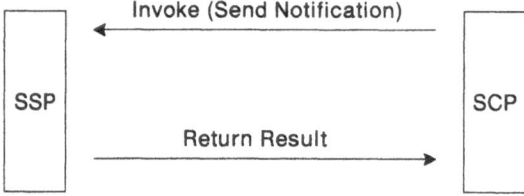

Processing Functions
1. Process the termination information data of the received TCAP input message.
   If the call ends normally, calculate the time elapsed between call connect and
   disconnect. If the call is abandoned (an error has occurred):
   - Calculate the time that has elapsed between call connect and the error.
   - Create an error message.
   - Issue a write command to the ERS error file, which is a VSAM ESDS file.
     The CICS Exec Link facility is used to access the error file. The
     application platform data manager controls and monitors the error file.
   - After completion of error the file write, check and analyze the delivered
     return code.
2. Update counters used for traffic measurements and statistics.
3. Free the main storage resources.

SSP-SCP Return Error
The SSP could not complete an ERS call with the message received from the SCP
after an initial inquiry. The message from the SCP requested "Sending Termination
Information". The SSP sends a TCAP Return Error message to the SCP.

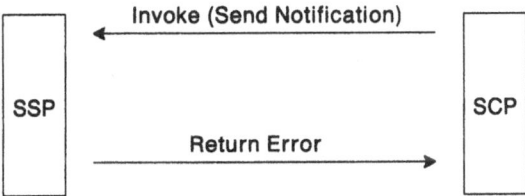

Processing Functions
1. Process the error information data of the TCAP input message of an
   abnormally ended call. The error code must be analyzed, and may contain the
   following information:
   - Unavailable network resource
   - Unavailable data
   - Unexpected data value (analyze problem data)
   - Protocol error (problem code from the SSP)
   - User error code.
2. Build error message and command to write the error message on an ERS error
   file. Following is a summary of the information to be provided in the
   command:
   - ERS error file name
   - Address of the field containing the error message
   - Access parameters (for example, type of access)
   - Return code area.

3. Issue a write command to the ERS error file, which is a VSAM ESDS file. The CICS Exec Link facility is used to access the error file. The application platform data manager controls and monitors the error file.
4. After completion of the error file write, check and analyze the delivered return code.
5. Depending on the TCAP error data and the return code, create a warning message to be sent to the Telco ERS operator.
6. Write the warning message on an application output queue. A CICS Temporary Storage Write command must be issued to put the message in the queue. There should be a particular ERS application output queue to receive all Telco ERS operator messages.
7. Update the counters used for traffic measurements and statistics.
8. Free the main storage resources.

SSP-SCP Reject

The SSP has found a protocol error in a TCAP response message received from the SCP after an initial inquiry message. The SSP provides information to be included in the reject message (such as responding transaction identification, correlation identification). The SSP sends a TCAP Reject message to the SCP.

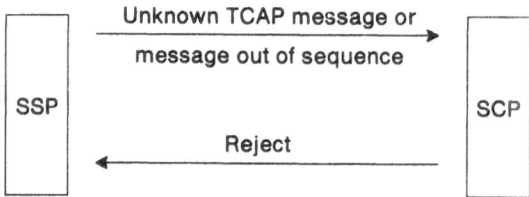

Processing Functions

1. Check and validate the data element of the TCAP message of a protocol error.
2. Analyze the protocol error code, and process the error data.
3. Create an error message and a command to write the error message on an ERS error file. Include the following information in the write command:
   - ERS error file name
   - Address of the field containing the error message
   - Access parameters
   - Return code area.
4. Issue a write command to the ERS error file, which is a VSAM ESDS file. The CICS Exec Link facility is used to access the error file. The application platform data manager controls and monitors the error file.
5. After completion of the error file write, check and analyze the delivered return code.
6. Depending on the TCAP error data and the return code, create an error message to be sent to the Telco ERS responsible operator.

7. Write the warning message on an application output queue. A CICS Temporary Storage Write command must be issued to put the message on the queue. There should be a particular ERS application output queue to receive all Telco ERS operator messages.
8. Update counters used for traffic measurements and statistics.
9. Free the main storage resources.

SCP-SSP Automatic Call Gapping

An asynchronous CICS transaction is periodically started to measure the ERS application load. The parameters for load measurement are:

- Transaction response time
- Values of the traffic counters
- ERS database load
- Mass calling on a PSAP destination number
- Queue length of unprocessed application requests.

If the load measurement signals an overload, an ACG is initiated. This can also be done by a command issued by the ERS operator personnel. In any case the ACG information will be sent to the SCP, embedded in a normal SCP-response message.

```
          Automatic Call Gapping (ACG)
        ◄─────────────────────────────
          The ACG information is
          included in the TCAP messages:
  SSP     Invoke (Connection Control),    SCP
          Invoke (Send Notification),
          Return Error
```

Processing Functions

1. Store the traffic counters from the last measurement period and determine the values of the traffic counters (number of ERS messages) for this period.
2. Analyze the database load from this time period.
3. Analyze the application input queue for unprocessed ERS application requests.
4. Analyze the calls for mass calling to a destination PSAP number.
5. Analyze the response time counters.
6. Check the number of active application clone tasks for the ERS application.
7. Process these parameters to determine the load.
8. If:
   a. there is a normal load situation, no action is required.
   b. there is an overload situation, or ACG initiated by a Telco operator, the overload level and sublevel will be determined by a table. Input values for this table are the referenced parameters, or the data delivered in the operator message. Output of the table processing is the level and sublevel of the overload situation.

9. Create and format the data element of the TCAP message containing the following ACG information:
   - Component type (Invoke)
   - Operation code (ACG)
   - Digits
   - Automatic call gapping indicators.
10. Update the counters for ACG control and statistics.
11. Call TCAP output message creating function to embed the ACG information in the next available response output message.

10.5 Administrative Units in SCP

There are three types of administrative units:
1. SMS Input Message Handling
2. SMS Message Processing
3. SMS Output Message Creating.

Because the functions involved in SMS input and output message handling are very similar for ERS SMS messages, only one administrative unit is described for "SMS Input Message Handling" and "SMS Output Message Creating".

The relationship between the administrative units and their interface to the Application Platform are shown in Figure 37.

All functional units use common Application Platform interfaces, access the database in much the same way, and work on closely related input and output data. Their common capabilities are:

1. **Interfaces to Application Platform and Queues**
   - CICS Temporary Storage Read Next command to read input message from the ERS application's update queue.
   - CICS Temporary Storage Write command to write output message on the ERS application's SMS output queue.
   - CICS Exec Link facility to update an ERS database record and write the ERS update application journal, and the ERS error file.

2. **Database Access**
   - Update a subscriber record via keyed access on the ERS database, which is a VSAM KSDS data file. To improve performance, parts of the file or the whole file may be kept in main storage. The application platform controls the storage method.
   - The platform must also ensure proper updates, where there are several copies of the ERS database reproduced on different SCP processors.
   - For improved performance, an update transaction may consist of a batch of updates affecting more than one database record.
   - A record is written to the ERS journal file, or the ERS error file, which are both VSAM ESDS data files.

3. **Input and Output Data**
   - The SMS-SCP header of the incoming update or retrieve report-data transaction contains information on the actions required by the ERS update application program.
   - The data part of the update message consists of subscriber data to be added, changed or deleted.
   - The data part of the update response message includes information on ERS record updating.
   - The data part of the "retrieve report-data" input message defines the specific parameters required to create the correct report data.
   - The data part of the "retrieve report-data" response message consists of the requested report data.
   - The SMS-SCP header of the response message contains additional control information for the ERS application in the SMS.

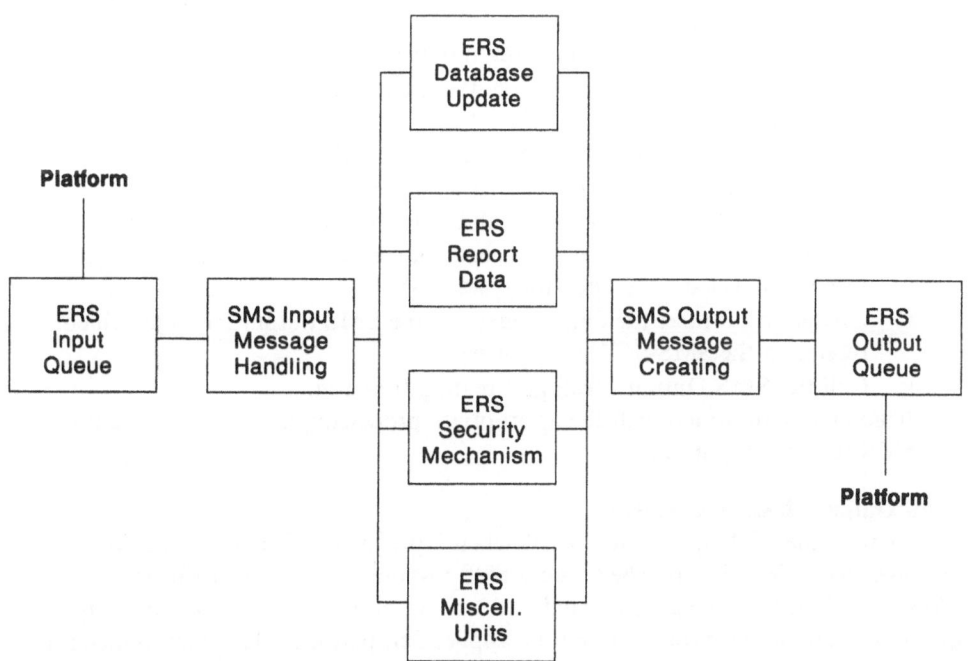

Figure 37. SCP Administrative Units and Application Platform Interfaces

SMS Input Message Handling
An SMS input message is received by the SCP, processed by the application
platform and written on an ERS application input queue, or on the ERS
application update queue in the case of an update transaction. The handling of an
application SMS input message starts with reading the message from the queue. The
message is then interpreted and analyzed. If there are no SMS-SCP protocol errors,
the data element of the message is given to the next application unit, depending on
the message type.

Processing Functions
1. Read message from ERS application input or update queue. A CICS
   Temporary Storage Read Next command must be issued to receive the message.
   The queues accessed by this command are allocated to the ERS application and
   used only for incoming SMS messages.
2. Get the main storage resources.
3. Analyze and process SMS-SCP header of incoming message.
4. Save SMS-SCP header information.
5. If an SMS-SCP protocol error was found in the analysis process:
   - Create an error message and a command to write the error message on an
     ERS error file. Include the following information in the write command:
     - ERS error file name
     - Address of the field containing the error message
     - Access parameters
     - Return code area.
   - Issue a write command to the ERS error file, which is a VSAM ESDS file.
     The CICS Exec Link facility is used to access the error file. The
     application platform controls and monitors the error file.
   - After completion of the error file write, check and analyze the delivered
     return code.
   - Create a warning message to be sent to the Telco ERS operator.
   - Write the warning message on an ERS application output queue. A CICS
     Temporary Storage Write command must be issued to put the message on
     the queue. (There is probably a particular ERS application output queue to
     receive all Telco ERS operator messages).
   - Create and format the data element of the SMS output message, which
     relates to the SMS-SCP protocol error.
   - Call the SMS Output Message Creating function.
6. If no errors are found, call the appropriate processing unit and provide it with
   SMS received input data.

SMS Output Message Creating
This part of the ERS application starts when a processing function, initiated by a
message from SMS, has finished, and a SCP response message must be sent to the
SMS. The output message is created and written on the ERS application output
queue for subsequent processing by the application platform data communications
part.

Processing Functions
1. Receive the output data supplied by the processing unit.
2. Restore SMS-SCP header control information. Determine and format the final SMS-SCP output message header.
3. Create an output message consisting of SMS-SCP header and application specific data.
4. Write an output message on a ERS application SMS output queue. A CICS Temporary Storage Write command must be issued to put the message on the queue. The queue used by this command is allocated to the ERS application and only used for outgoing messages to the SMS.
5. Update the counters used for traffic measurements and statistics.
6. Write relevant information, collected during the ERS logic process, on an ERS application journal. The journal is a VSAM ESDS data file and is written by use of the CICS Exec Link facility. The application platform controls and monitors the ERS journal.
7. Free the main storage resources.

SMS-SCP Update of ERS Database
An update transaction received in the SCP from the SMS could make a change, addition, or deletion of an ERS database record. An update can be a single transaction or a batch of transactions referring to more than one database record. The update process is a logical unit of work; If it fails, all the changes performed by this logical unit of work will be backed out by the application platform using dynamic transaction backout or emergency restart.

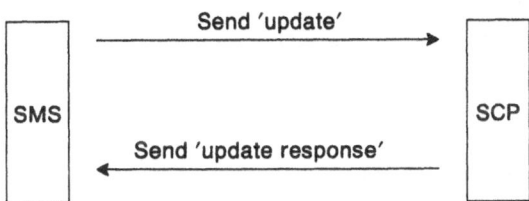

Processing Functions
1. Check, validate, and edit input data of the update message.
2. Update counters used for traffic measurements and statistics.
3. Create a command and build control blocks for record update on ERS database (file name, key, Input/Output area, access parameters, return code area etc.).
4. Update a database (the actual update of all ERS database copies on the SCP processors is performed and controlled by the application platform).
5. After completion of the database update, check and analyze the delivered return code.
6. If:
   a. The update was successful, create the response "update successful".

   b.   The update was not successful, write the update record on the ERS error
        file, and create the response "update not successful".
7.   Create and format the final data element part of the update response message
     to the SMS.
8.   Call the SMS Output Message Creating function.

SMS-SCP Retrieve Report Data for ERS
The ERS application maintains measurement data. The SMS polls the SCP node at
fixed times to obtain report data. The SCP can generate reports daily, hourly, or
every five minutes. The name of the reports, schedules, an so on are maintained in a
parameter control file.

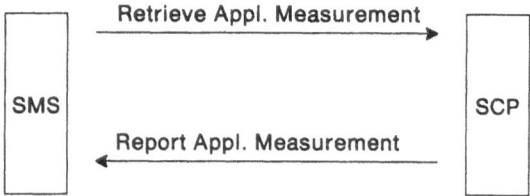

Processing Functions
1.   Analyze and process the message input data element.
2.   Read the ERS application measurement counters.
3.   Create an ERS application report message, containing:
     •   Application identification
     •   Indication that data is sent for one or more interval
     •   Start and finish time of the interval
     •   Data part:
         –   Number of successful responses
         –   Number of response messages with errors
         –   Number of misrouted queries
         –   Number of queries rejected.
4.   Create and format the data part of the response transaction to be sent to the
     SMS.
5.   Call the SMS Output Message Creating function.

Part 5. Private Virtual Network

This part provides detailed information about the Private Virtual Network (PVN). It examines what PVN is, how it operates in the IN, and the factors to consider when implementing PVN.

This part is PVN-specific; it adds to, and does not replace "Part 1. Overview of the Intelligent Network" of this report.

Chapter 11. PVN Service Description

11.1 Overview

Basic Service

The Private Virtual Network Service is designed to provide the functions of private networks using the Public Switched Telephone Network (PSTN).

It offers the user dynamic allocation of network resources and a uniform numbering plan over geographically independent locations. The major characteristics are:

- Originating and terminating screening
- On-net (using the virtual facilities of the network)
  An on-net location is one subscribing to PVN. It has an abbreviated dialing facility according to a uniform numbering plan. On-net locations may be served by dedicated or dial access arrangements.
- Off-net
  An off-net location does not subscribe to the PVN
- Uniform numbering plan
- Dynamic allocation of resources
  - Managing physical resources
  - Access points
  - Screening
- Customers control administration of their network.

The Private Virtual Network data is stored in a central database which is accessed via CCS7 when the PVN member originates a call or receives a call.

Background

Private telecommunications networks are used by most large companies, government, and other organizations. Leased line voice and data services have been used extensively to reduce communications costs while improving security and quality. Private services include:

- (voice) private tie-lines
- (data) 1200 - 9600 bps leased line services
- (data) high-speed 56, 64 kbps leased line services
- (data) broad-band leased line services.

Companies with private networks to support these and other services must plan, manage, and maintain their networks.

Telecommunications technology will facilitate the cost-effective delivery of services such as high-speed circuit, switched data, packet switched data, video transmission, and integrated or shared-use service capabilities.

The **Private Virtual Network** service offers the function of a private network, but (because the service is provided over the PSTN) using shared facilities. This means the service may be available at a significantly reduced cost. The organization only pays for the specific service used rather than paying for facilities which have been leased, or which are for private use. Moreover, as providing services is the network operator's responsibility with PVN, organizations do not require the planning, maintenance, and management staff necessary with private networks.

Benefits
The potential benefits are:
- Service User
  - Convenience
  - Speed and ease-of-use
  - Multiple location access
  - Uniform dialing
  - Availability to company network.
- Service subscriber
  - Cost savings (PSTN) can be shared with other users
  - Pass network management responsibility (planning, installation, and maintenance) to the network operator
  - Flexibility to adjust to changing needs the allocation of network facilities
  - Reduce capital investment
  - Improves business communication (productivity and efficiency)
  - Potentially improve security or reliability
  - Accounting for usage
  - Benefit from Telco deployment of new technology without incurring additional expenditure.
- Network Operator
  - Increase Revenue by more rapid provision of service
  - Improve customer service by faster deployment of resources
  - Protect market share
  - Optimize network utilization and investment
  - Facilitate maintenance of all network operations.

11.2 Functional Description

General

PVN service is implemented using databases provided by the network operator. They are located at Service Control Points (SCPs), and called the Business Service Databases (BSDB). The PVN service may be used in the following way:

1. A PVN call is recognized when the user dials a Service Access Code (SAC), or by line identification when the customer premises equipment (CPE) goes off-hook, or a remote access user dials a specific number.

2. The exchange recognizes the request for PVN service and the service request is forwarded to the nearest properly equipped SSP. The SSP suspends normal call processing and sends a query to the SCP.
   Note: If the local exchange does not have SSP capability, some line features may not be available to users.
   The SCP interrogates the PVN database to find call handling instructions for the specific ANI, line, and access code dialed and/or Personal Identification Number (PIN). Queries and responses between the SSP and the SCP use CCS7 protocol.

3. The response message from the SCP indicates whether or not the call should be completed, how it should be routed, and what information should be in the AMA record for the call. The response may also instruct the SSP to obtain additional input from the caller, such as an authorization code.

4. Call set-up instructions are sent to the appropriate network resources to set up the call.

Services

Note: This section is based on *The Federal Telecommunications Systems (FTS 2000) Procurement Document Reference*, KET-JW-87-02, Amendment No. 1, March 31, 1987.
The following PVN services are described in this chapter:

- Switched service for the transmission of voices or data at up to 4.8K bps
- Switched data service for the transmission of data in digital format at 56K bps and 64K bps
- Switched digital integrated service for the digital transmission of voice, data, image, and video.

On-Net: PSTN access: An on-net location is a location subscribing to PVN. It has an abbreviated dialing facility according to a uniform numbering plan. On-net locations may be served by dedicated or dial access arrangements.

Virtual On-Net location: Dial access is access to PVN using some portion of the public switched network (hereinafter referred to as "remote access"). Remote access locations are referred to as "virtual on-net" locations and have public switched network numbers and abbreviated dialing numbers.

In the future, on-net locations should be able to call virtual on-net locations by dialing the abbreviated number.

Off-net location: An off-net location does not subscribe to the PVN, and is also known as remote access.

The PVN switched voice service supports calling types which connect originating and terminating locations as follows:

a. On-net to on-net.
b. On-net to virtual on-net.
c. On-net to off-net.
d. Virtual on-net to on-net.
e. Virtual on-net to virtual on-net.
f. Virtual on-net to off-net.
g. Off-net to on-net (including travel access).
    For example, a travelling salesman calls his company to report sales information. His call is forwarded by the PVN system to an on-net location.
h. Off-net to virtual on-net (including travel access).
i. Off-net to off-net (including travel access).
j. Connection to and from physical private facilities.

Calling types (b), (d), (e), (f), and (h) include Virtual On-Net Calling.

While off-net calls can be made by public callers, the requirements for off-net calling types stated below only pertain to users at temporary locations.

Virtual On-Net Calling: allows users at virtual on-net locations, equipped with either DTMF or dial-pulse telephones or ISDN equipment, to call on-net, virtual on-net, and off-net locations.

Off-Net Travel Calling: allows authorized persons located at temporary locations to call on-net, virtual on-net, and off-net locations. The user must enter a valid authorization code before the call is completed.

Service Delivery Points: PVN provides services and features between service delivery points, each of which is defined as the combined physical, electrical and service interface between the PVN and customer premises equipment, off-premises switching and transmission equipment, and other services (such as those provided by Centrex and telephone central offices). Service delivery points serve on-net and off-net locations and may be located on or off customer premises.

An on-premises service delivery point is normally located at the customer premise.

Switched Voice Service: The switched voice service supports connections for voice or analog data up to (9.6 kbps) and allows calls initiated from PVN on-net locations to be connected to all on-net and off-net locations by direct station-to-station dialing. Calls may be initiated at off-net locations through direct station-to-station dialing.

The switched voice service interconnects the following types of terminal equipment:

• Single-line telephones
• Multiline key telephone systems
• Electromechanical, analog and digital private branch exchanges (PBXs), Basic Access and, Primary Access ISDN (CPE)

- Centrex stations
- Data circuit terminating equipment
- T1 digital terminating equipment
- CCITT Group I, II and III facsimile apparatus
- Secure voice and secure data equipment
- Other equipment (usually on customer premises) used for connection to public and private switched networks.

The switched voice service also provides network interconnectivity between PVN and the public switched network.

The service requires a uniform numbering plan whereby each station receiving PVN service is identified by a unique on-net directory number.

PVN provides network intercepts or recorded announcements when a call or call attempt cannot be completed. These are provided for the following conditions:

- Number disconnected
- Number reassigned
- Partial dial
- Incorrect number of digits dialed
- Time out during dialing
- Network congestion
- Preemption of access due to the provision of service for high priority authorized users
- Denial of access to off-net and international calls
- Denial of access to features.

Recorded Message Announcements: Allows authorized personnel to record message announcements within the network. The recording shall be assigned an on-net number, and shall be accessible by both on-net and off-net stations. The maximum message announcement length shall be three minutes. A call to the announcement must be answered within five rings, and barge-in access to the announcement shall be permitted. No call shall remain connected for more than a partial plus one complete iteration of the message.

Attendants: Will be available constantly, so that PVN users will not receive a busy tone when calling for attendant services. Calls for attendant-service should be answered within five rings in 90% of cases. Attendants provide the following services:

- Where automatic network audio conferencing is not possible, set up conferences and connect all conferees.
- Assist users with dialing difficulties, and remain on the line until the call has been completed.
- Verify authorization codes given verbally by callers accessing PVN from off-net and on-net locations. Once a code has been verified, the attendant proceeds to set up the call requested by the caller.
- A PVN user can call a PVN attendant by dialing a special number to report a stolen or lost authorization code, or to obtain a credit adjustment for a

completed call to a wrong number, interruptions of calls, or unsatisfactory transmission.

Authorization Codes: Provide calling identification and class of service for PVN users. If identification cannot be made by other means, the same authorization code can be used for originating on and off-net calls for billing and class of service purposes. The same code can also be used to override service restrictions.

Service derived from an authorization code has priority over that derived from other means. Requirements for and definitions of authorization codes are specified by the service subscriber.

The format of the authorization code is determined by the Network Operator. The dialing sequence must be such that the PVN network is alerted that an authorization code will be entered, in order to minimize the processing delay of calls not requiring the use of that feature.

Potential Fraud: Repeated attempts to use the same or different invalid authorization codes, or other identifiable patterns, may be a deliberate effort to penetrate PVN by unauthorized individuals. The Network Operator implements procedures for detecting what appears to be fraud, and promptly reports these cases to the appropriate authority.

Call Screening (Closed User Group): is a set of features that determine a call's eligibility to be completed, based upon class of service information associated with the user, the station, or the trunk group.

Class of Service: is determined from ANI, authorization code, traveling classmark, or trunk group. The class of service derived from an authorization code takes priority over that from other means. Service classes identify access and feature restrictions:

- Access restrictions include access to off-net calling, access to portions of the PVN, and access to other than specified numbers.
- Feature restrictions allow or restrict access to network features by users, or groups of users.

Time of day and day of the week or year restrictions are possible. These would be implemented on a per station, per location, and per authorization code basis. This type of restriction can be used to prevent unauthorized use of PVN.

Class of Service Override: Allows a PVN user to override the terminal equipment's preassigned class of service treatment for the duration of the call by entering a valid authorization code. The records of call details must reflect all relevant call data to be charged to the authorization code user rather than to the origination station.

Traveling Classmark: Provides for the acceptance of traveling classmarks from any PVN locations served by PBXs and Centrex that are able to provide these classmarks. User calling characteristics and limitations may be determined from the traveling classmark.

This feature will be available to all locations served by PBXs and Centrex equipped with the traveling classmark capability, and all types of representative terminal equipment.

Code Block: This prevents ineligible users, stations and trunks with certain class of service access restrictions from calling specified area codes, exchange codes and countries, by translating the digits dialed and comparing the results with the class of service. Blocked calls are intercepted and connected to appropriate network recorded announcements.

Network Audio Conferencing: Conference call support is provided for a mixture of on-net, off-net, and virtual on-net stations, and includes the following conference features:

- "Meet-Me"—allows a number of users to be connected in conference by dialing an access code at a predetermined time or as directed by the PVN attendant.
- Present—allows users authorized to place such calls to arrange conferences using lists of conferees.
- Add-on—allows the user (who is the call controller) to add conferees to an existing conference.
- Attendant-Assisted—used by callers who cannot initiate a conference call from their telephones.

On-Net to Off-Net Translation: translates an abbreviated on-net number to a public switched network number (off-net location). Time of day and week, office status (whether open or closed) and other factors effect translation.

Off-Net to On-Net Translation: routes calls that were dialed using the PSTN dialing to on-net locations. For example, the feature can be used to temporarily reroute a call to an on-net location that has just been added to the PVN network, but is still being called at the old off-net number.

Off-net to on-net translation will be made available to all types of representative terminal equipment.

Inward Station Access: allows off-net callers to be connected to predesignated PVN stations by dialing certain customer GNS or other GNS numbers.

Inward Selected Access: allows off-net callers to be rerouted to network locations by dialing specified GNS numbers. Once the caller has been connected, the network provides a recorded announcement identifying the DTMF selection digit(s) that provide further routing of the call.

For example, if a taxpayer wants information from an Internal Revenue Service (IRS) Regional Information Center, he dials the IRS commercial free number, and receives a recorded announcement instructing him to "press two to ask for tax forms, four to file late tax returns, six to register complaints against tax system loopholes", and so on.

Switched Data Service: The switched data service provides a synchronous, duplex, totally digital, Service Delivery Point-to-Service Delivery Point circuit-switched service at 56 kbps and 64 kbps. It is used to support workstations, host computers, personal computers, terminals and other office support equipment.

The network operator provides a numbering plan for service users to use when they call each other. Calling capability is spontaneous and does not require scheduling. The service provides the data terminal with network-derived clocking.

This service is delivered directly to data terminal equipment or indirectly to data terminal equipment through a digital PBX.

An authorization code can be used to identify the caller, device or application program for purposes of call screening, indicating the requested feature, or billing (especially when ANI is not available). The network operator treats authorization codes for this service in a manner similar to that specified for switched voice service.

Switched Digital Integrated Service (Narrow Band ISDN): Is used to integrate voice, data, image and video services by digital connectivity to PVN user-equipment. The network operator provides two digital interfaces under switched digital integrated service: ISDN and T1.

Service User's Perspective
Note: The rest of 11.2, "Functional Description" on page 202 is an extract from *Additional Service Switching Point Capabilities (including Private Virtual Network Services)*, Bellcore TA-TSY-000402.

PVN Calls: A PBX or Business Group member usually makes a PVN call by dialing a prefix (for example by dialing an "8"). If a PBX is served by a common trunk group, a special prefix (or SAC) "XX code" could be generated and outpulsed by the PBX to indicate a PVN call. The XX code could also be used by an individual line to indicate a PVN call.

A caller may make either on-net calls or off-net calls with PVN service. On-net calls may consist of various numbers of digits according to the agreed numbering plan.

On some calls, the caller may be instructed to input an authorization code in order for the call to be completed. If the caller is authorized to make the call, he would normally enter the proper authorization code by means of Dual-Tone Multifrequency (DTMF) digits or Rotary dial digits.

The Remote Access to PVN feature permits a caller to dial a directory number from any station in order to be connected to PVN service. The call will be routed to an SSP switch providing remote access for the particular PVN. The SSP will provide an announcement or recall dial tone to the caller who will then dial a PIN. Upon receipt of another announcement or recall dial tone from the SSP, the caller will be able to place a call on the PVN network in the same manner as though he had accessed the PVN network from their regular business station.

A PVN customer may provide an attendant to assist rotary dial customers to input authorization codes and to enter the necessary information on remote access calls.

Extension Calling: All intralocation and interlocation calls within a PVN Area Wide Business Group may be placed without a PVN access code by dialing an extension number typically consisting of four digits (if the Business Group involves fewer than approximately 8000 lines) or five digits (for a Business Group having more than approximately 8000 lines). The number of lines that can be handled with four digits will depend upon the number of prefix codes (such as an 8 or 9) used for other purposes. In many cases the limit may be 7000 or 8000 lines. Interlocation extension calls may be completed over PVN facilities.

Callers from outside the PVN Business Group access PVN lines with a normal public dialing number which appears no different from standard telephony addresses. The final digits of the public dialing number are usually the same as the extension code used by callers within the Business Group to access the concerned line.

Service Subscriber's Perspective

1. Private Virtual Network Service is intended to provide the same or better service than Private Network Service - at equal or less cost.
2. The Network Operator, rather than the Service Subscriber, is responsible for planning, installation, maintenance, administration of the network.
3. Service changes may be provided on demand, (immediately) rather than as a result of a traditional purchase order process.
4. The Service Subscriber maintains the service control, but is not burdened with the responsibility for the operation of his network.

Network Operator's Perspective

(1) PVN Access
Dedicated PVN Line
The Network Operator should be able to assign certain lines as dedicated PVN lines. All calls originating on such lines should be treated as PVN calls without the need for identification via access codes.

Dedicated PVN Trunks
The Network Operator should be able to assign certain trunks (for example, PBX trunks or trunks from an end office) as dedicated PVN trunks. All calls originating on such trunks should be treated as PVN calls without the need for identification via access codes.

Business Group Lines
All calls dialed by a Business Group line with a prefix that has been assigned to indicate a PVN call for that Business Group customer should be treated as PVN calls. Provision should also be made on calls using Automatic Flexible Routing (AFR) arrangements (for example, 9 + calls) to specify that a call not originated as a PVN call should be treated as a PVN call.

Non-dedicated PVN Lines
It is planned to assign a Service Code (SC) that may be used on certain lines (for example, PBX lines) to access a PVN. Translations should be provided to indicate if this code should be accepted from a line.

Direct Connect Line
The Network Operator should be able to assign certain lines within a Business Group as PVN Direct Connect Lines. All originations on such lines should be treated as PVN calls.

The above capabilities should be provided in such a manner that they are not limited to PVN calls. That is, a "trigger" table should be provided that would permit identification of calls from lines or trunks with certain classes of service (for example, PVN) or dialed with certain prefixes (for example, an "8" from a Business Group station or a XX code from a POTS line) as calls requiring database queries.

This table should also contain the special indicators required to be included in the query message for each type of call. Examples of such indicators are ANI information digit pairs and subsystem numbers.

A member of a PVN may originate a PVN call while connected to another party (for example using the add-on or transfer features). When this occurs, the SSP should make a regular PVN query to the SCP and use the information in the response message to establish the third leg of the call.

(2) PVN Dialing Plan

After a customer has indicated that they are placing a PVN call as described above, they should be able to complete their call by using a predetermined dialing plan. The Network Operator should be able to specify for each PVN subscriber, the dialing plan to be used, and the number of digits to be dialed.

If the digits received on a PVN call do not agree with the dialing procedure, the call should be blocked at the PVN switch and reorder tone or reorder announcement returned.

Incoming Business Group tie-trunks may carry PVN calls. The Network Operator should be able to specify for each incoming tie-trunk whether all calls on the tie-trunk should be treated as PVN calls or if the calls received should be treated the same as calls originating from stations within the Business Group. Calls over certain tie-trunks may include a private network traveling class mark. The Network Operator should be able to specify incoming PVN access tie-trunks on which the final one or two digits received is a private network traveling class mark. If a traveling class mark is received, it should be included in the query message to the SCP. The switch should be capable of receiving rotary dial, DTMF and MF signaling on incoming PVN tie-trunks.

A "station group designator" may be received on a direct trunk from a PBX. Station group designators are identical to the traveling class marks described above and should be treated in the same manner.

(3) Remote Access Dialing Plan

A member of a PVN should be able to place a call from any remote station to gain access to their PVN. Provision should be made for the Network Operator to assign a number or numbers to a PVN customer to be used for this purpose. Upon receipt at an SSP of a remote access number, the call should be handled in one of the following ways:

- The Network Operator should be able to indicate that calls to certain directory numbers should result in a query being made to an SCP database. Translations should be provided that would indicate for each of these directory numbers, the ANI information digits and the subsystem numbers to be included in the query message. This capability would permit the Network Operator to provide other database services to be reached by the dialing of a directory number.
  For PVN service, the SCP would then request that an announcement or recall dial tone be connected and that a specific number of digits be collected from the caller as a PIN. (A PIN is used to identify the caller or the caller's capabilities). The message from the SCP would also specify that a second

announcement or recall dial tone be connected and the normal PVN dialing sequence for the PVN customer be collected. A second message should then be sent from the SSP to the SCP reporting the digits received. The SCP will respond to this message with routing and billing information for the call.

- If a particular type of switch presently has a remote access capability, this capability could be modified to provide remote access to PVN. With such an arrangement, upon receipt of a PVN remote access number, the SSP without making a database query should return an announcement or recall dial tone to the caller who would enter a PIN. The Network Operator should be able to specify for each PVN customer the number of digits to be collected as a PIN. Upon receipt of a PIN of the proper length, the SSP should connect another announcement or recall dial tone to the caller. The caller should then be able to dial a PVN call as though they were placing the call from a station that was part of the PVN. A query should then be made to the SCP to determine if the PIN is correct, and to obtain routing and billing information for the call.

If the caller should enter an incorrect PIN, the SCP may send another message to the SSP requesting it to connect an announcement or recall dial tone and to collect a specific number of digits as a PIN. Upon receipt of these digits from the caller, the SSP should send another message to the SCP transmitting the digits received. The number of times that a caller may dial an incorrect PIN without being disconnected is included in the PVN customer's data in the SCP. When this number is exceeded, the SCP will instruct the SSP to abort the call.

The SSP should be arranged to accept the DTMF signal "*" while receiving either the PIN or the normal PVN dialing sequence on a remote access call. The DTMF signal "*" is used by the caller to indicate that they have made a mistake. When the "*" signal is received, the SSP should discard the digits already received and be prepared to receive the entire sequence of digits.

The SSP should return answer supervision towards the originating office when connecting the first announcement or the first recall dial tone to the caller. At least one cycle of audible ringing should be returned to the caller before the start of the announcement or the recall dial tone.

Non-SSP PVN Switches (Extensions of PVN): The capability should be provided in PVN switches not equipped as SSPs to recognize calls as PVN calls. A non-SSP PVN switch should also be able to provide the dialing patterns as necessary.

PVN Offices not equipped as SSPs should route all PVN calls to an SSP.

(5) PVN Calls from PBXs Using MF Signaling

An SSP should be able to receive calls using MF Signaling from PBXs so equipped. The switch should be able to accept normal signaling on these trunk groups.

(6) Extension Calling
The Network Operator should be able to specify for each Area Wide business group, the digits to be used to indicate intra-switch calls and to indicate calls to each of the other locations of the customer. For calls to other switches, the Network Operator should be able to specify prefix digits to be outpulsed, and the routes to be used for each remote location. For PVN stations served by an SCP, the Network Operator should be able to specify that on calls to certain locations, a query be made to the Business Service Database (BSDB) for routing information. (The BSDB is the SCP database that handles PVN calls). The Network Operator should also be able to specify that on calls to numbers assigned as Network Number Calling numbers, a query be made to the BSDB for routing information.

(7) SCP Queries
The SSP should formulate and send a CCS7 query to the SCP under the following conditions:
1. Dialing has been completed on a PVN call from a station or tie line served by the SSP
2. A PVN call has been received from a non-SSP switch
3. A PVN call has been received via remote access
4. A PVN call has been received via remote access and the caller has completed dialing their PIN and their PVN dialing sequence.
5. An extension call requiring database routing information has been received.

(8) Completed as Requested
The actions to be taken on a given call depend on the response message received from the SCP. This response should include information needed to route the call and may request the SSP to send additional information, such as whether the call was answered and, if so, the call duration.

The response message format should be similar to that described for the query message. Because connectionless service is used, the response will be contained in an SCCP unidata message. When the SCP requests the SSP to send termination information, the response message from the SCP should contain the Transaction ID assigned by the SSP.

The response should contain an INVOKE Component which instructs the SSP to route the call based on the value of the data elements in the message. The following actions may be requested in the response message:
A. Route the call over the public network
B. Connect an announcement to the calling station
C. Route the call over private facilities
If an AMA record should be made, a Billing Indicators field will be included in the response message.

(9) Authorization Codes

Instead of sending a Response as a result of a query message, the SCP may request that an authorization code be obtained from the caller. This is done by sending a "Conversation with Permission" message with an INVOKE "Play Announcement and Collect Digits".

This feature should be provided in such a way that its use is not restricted to PVN service. That is, for any service on which the SSP has sent a query and is waiting for a response, it should be able to accept an INVOKE "Play Announcement and Collect Digits" and to forward the digits collected to the SCP.

Upon receipt of this message, the SSP should make the necessary connections to play the requested announcement to the caller and to accept DTMF digits from the customer. If this capability is normally provided by means of an optional separate intelligent peripheral, the SSP should also have the capability to play announcements and collect digits without the use of the intelligent peripheral. If the calling station is connected directly to the SSP, the SSP should also be prepared to accept rotary dial digits from the caller. The number of digits to be collected will be specified in a "Number of Digits" field.

Receipt of the first digit from the caller should result in the removal of the announcement. If the switch does not have the capability to detect a digit while playing an announcement, it should play the announcement once and then connect recall dial tone and be prepared to detect digits. This procedure will permit the use of an announcement such as "At the tone please enter your authorization code".

In many cases, the announcement specified in the message from the SCP will be "recall dial tone". In this case, recall dial tone should be connected in the normal manner.

Upon receipt of the authorization code from the caller, the SSP should send a RETURN RESULT in a "Conversation With Permission" message to the SCP. This RETURN RESULT should indicate "Caller Interaction" and include the digits input by the caller. The SSP should also restart the timer that was used when the original query message was sent to the SCP.

If the caller should remain off-hook for five seconds and dial no digits, the SSP should respond as described in the previous paragraph. However, the "Digits Length" field will have the value of zero, indicating that no digits were received. This will permit the SCP to return routing instructions to the SSP to route the call to the PVN customer's attendant. The attendant could then set up a new call and enter all the information required to complete the call.

If the caller should remain off-hook and dial insufficient digits, but at least one digit, the SSP should send a RETURN RESULT as described above. This RETURN RESULT should include the digits input by the customer. This will permit the SCP to request the customer to try again or to send instructions to abort the call. Normal partial dial timing should be used to detect this condition.

If the caller should abandon without dialing an authorization code, the SSP should send a Response message to the SCP with a RETURN RESULT including the Standard User Error Code "Caller Abandon".

The SSP should be arranged to accept the DTMF signal "*" while receiving an authorization code. The DTMF signal "*" is used by the caller to indicate that they have made a mistake. When the "*" signal is received, the SSP should discard the digits already received and be prepared to receive the entire sequence of digits.

If there is a protocol error in the message from the SCP, the SSP should respond with an error message with "REJECT" as its Component Type Identifier. For other errors in the SCP message, the error message sent by the SSP should indicate "RETURN ERROR". The SSP should also connect reorder tone or reorder announcement to the calling party when an error condition occurs.

On calls originating in non-SSP end offices, the SSP should return answer supervision to the originating office before connecting the announcement. If CCS7 signaling is used for call setup from the originating office, and a subsequent Address Complete Message and/or Answer Message is received from the far end, an Address Complete Message and/or a second Answer Message should not be sent to the originating office. If, instead of an answer, a Release Message with Cause is received, the SSP should connect the tone or announcement required for the particular cause.

(10) Remote Access Calls

If the arrangement described in Remote Access Dialing Plan (above) is provided for Remote Access, the SCP will not send a Response to the query message, but will request that additional information be obtained from the caller. This will be done in a manner similar to that described above in connection with authorization codes. However, the "Conversation With Permission" message will contain two INVOKES, each with the Operation Code "Play Announcement and Collect Digits". The first INVOKE will specify the number of digits to be entered as a PIN. The second INVOKE will contain 11111101 in the "Number of Digits" field. 11111101 in the "Number of Digits" field of the INVOKE, "Play Announcement and Collect Digits", indicates that the SSP should determine the number of digits to collect using the normal translation information for the particular PVN customer.

Upon receipt of this message, the SSP should connect the announcement and collect the number of digits specified in the first INVOKE. It should then connect the announcement specified in the second INVOKE and collect the digits that would normally be dialed by a member of the particular PVN on a PVN call. If the switch does not have the capability to detect a digit while playing an announcement, it should play the announcement once and then connect recall dial tone and be prepared to detect digits. After receiving both sets of digits from the caller, the SSP should send two RETURN RESULT components in a "Conversation with Permission" message to the SCP. Each RETURN RESULT should indicate "Caller Interaction" and include the numbers received from the caller in a digit field. When reporting the second set of digits, the Nature of Number field should indicate international, national, or network-specific depending upon the prefix dialed.

On calls originating in non-SSP end offices, the SSP should return answer supervision to the end office before connecting the announcement. If CCS7 signaling is used for call setup from the end office, the SSP should subsequently connect a tone or announcement on the call if a Release Message with Cause is received from the far end. If an Address Complete Message and/or an Answer

Message is received from the far end, the SSP should not repeat them to the originating office.

If the caller should remain off-hook for five seconds and dial no digits, the SSP should send a single RETURN RESULT component in a "Conversation With Permission" message to the SCP. The "Digits Length" field should have the value of zero, indicating that no digits were received. This will permit the SCP to instruct the SSP to route the call to the PVN customer's attendant to assist the caller. Similarly, if no digits are received from the caller by a switch that normally collects a PIN and a dialed number prior to sending a query, the switch should send a query with the "Digits Length" part of the PIN digits field having the value of zero.

If the caller has entered an incorrect PIN with either of the Remote Access call procedures, the SCP may send a message requesting the SSP to have the caller redial their PIN. This will be another "Conversation With Permission" message containing a single INVOKE to "Play Announcement and Collect Digits". The SSP should connect the announcement and collect the number of digits requested. After receiving the digits, the SSP should send a RETURN RESULT in a "Conversation With Permission" message to the SCP. This RETURN RESULT should indicate "Caller Interaction" and include the numbers received from the caller in a digits field. If the PIN is still incorrect, the SCP will send a response message requesting that the SSP either collect the digits again or connect reorder tone or reorder announcement.

The SSP should be arranged to accept the DTMF signal "*" while receiving either the PIN or the normal PVN dialing sequence on a remote access call. The DTMF signal "*" is used by the caller to indicate that they have made a mistake. When the "*" signal is received, the SSP should discard the digits already received and be prepared to receive the entire sequence of digits.

(11) Resource Counters

It is expected that the SCP will contain resource counters for routes between switch locations for many PVN customers. When the SCP receives a query message for a PVN call, it will check the resource counter associated with the particular route. If the resource counter indicates that there is an available resource, it will increment the counter and send a response message to the SSP instructing it to complete the call. The response message will also request the SSP to send termination information at the end of the call so that the associated resource counter can be decremented.

Since it is possible that the response message carrying routing instructions or the termination message for a call involving a resource counter may be lost, means must be provided to verify that the resource counter values are correct.

In order to do this, the SCP will periodically send queries to the two SSPs associated with a particular resource. This query will request the SSPs to check the number of existing calls using the particular resource and to report this number back to the SCP. The SCP will then add the numbers received from the two SSPs, and correct the resource counter if necessary. This procedure would normally take place during periods of low traffic.

(12) Terminating Calls

A.   Calls Terminating to PVN Stations
      The Network Operator should be able to indicate in the switch translations for
      both SSP switches and non-SSP switches, PVN stations that are restricted
      from receiving non-PVN calls. The only calls to be completed to these lines
      are those calls received with a PVN indication.

B.   Hop-off Calls and Calls to Private Facilities
      Existing Business Group capabilities permit a call received over a private
      facility to hop-off using AFR lists or to be connected to other private
      facilities. These same capabilities should be provided for PVN calls received
      over public facilities when CCS7 signaling is used for call setup between the
      originating SSP and the terminating PVN switch.

Relation to ISDN Supplementary Services
The implementation of ISDN will have an impact on the PVN application, however
this has not been examined in detail. Some preliminary considerations are:
      ISDN complements PVN by providing standardized links and the
      D-channel instructions.
      D-channel instructions supersede some PVN application
      instructions.
      Additional PVN instructions may be required to take advantage of
      ISDN capability.

11.3 Standards

Service
The following documents describing PVN standards were the primary source of
information about PVN standards for this study:
      Bellcore recommendation TA-TSY-000402, *Additional Service Switching Point
      Capabilities*.
      Bellcore recommendation TA-TSY-000460, *Business Services Database (BSDB):
      SCP Applications Designed to Support Private Virtual Network Services*.

11.4 Service Interaction

Service User
The service user accesses PVN from POTS DTMF or ISDN CPE by dialing or keying an access code (for example ANI, SAC or a special number). PVN may respond by requesting a PIN, in which case the user must dial or key that in. When an error occurs, the service user is automatically connected to a network operator attendant, or he dials the attendant's number.

Guidance is provided by audio announcements triggered by the SCP or by a PVN attendant.

Service Subscriber
The Service Subscriber maintains the class of service, PIN, and other data for each service user through input to the SMS or other input to the network operator.

The SMS provides a user interface with a hierarchy of menus to support and guide the subscriber in the service related activities. interacting with the SMS, the service subscriber enters the following data:
- Security passwords and PIN
- Access instruments, characteristics (speeds, protocols)
- Uniform numbering plan
- Routing parameters
- Screening parameters
- Real number
- Recorded announcements
- Business group designations
- PVN trunks required.

Network Operator
The network operator administers and maintains service specific data and logic, as described in Chapter 2, "Functional Characteristics Common to Selected IN Services".

Access Instrument
The telecommunications equipment used to access switched voice service includes:
- Single-line telephone sets.
  Service for single-line sets is delivered to the PVN telecommunications point for interconnection of local exchange PVN carrier facilities to premise equipment.
- Multiline key telephone systems.
  For multiline key telephone systems, the PVN switched voice service is delivered to the telecommunications point for interconnection of local exchange carrier facilities to premise equipment.
- PBXs.
  PVN is delivered to the trunk side of the PBX.

- Centrex.
  PVN switched voice service is provided at the trunk side of Centrex; this allows a Centrex group to be part of the PVN.
- Data circuit-terminating equipment.
- CCITT Group I, II, and III facsimile apparatus for document transmission.
  When a CCITT Group I, II, or III facsimile apparatus interfaces directly with the network operator's facilities, the service delivery point is at the facsimile apparatus line interface.
- T1 digital terminating equipment.
- Secure voice and secure data equipment.
- Basic Access.

The service delivery points for switched voice service are defined for each type of interface to the representative terminal equipment listed above.

11.5 Billing

SSP

The network operator provides a system for collecting, aggregating and formatting all billing information for PVN services. Call details are identified to the lowest level of call origination, that is, the originating station's calling number, location number or authorization code.

Station identification is provided by Automatic Identification of Outward Dialing (AIOD), Automatic Number Identification (ANI) or an authorization code. Switched digital integrated service for ISDN must provide station identification as specified in the CCITT Recommendation Q.931.

Call details include:

- Calling number
- Date of message
- Time of message origination
- Message length and duration, measured in time increments and/or billing units
- Called number
- Called location
- PVN service code, indicating type of service and features used
- Usage charges.

SCP

Currently, the SCP does not record any billing information.

SMS

The SMS must generate information about the service subscriber's session activities (updates, number of queries, requests for reports) to the network operator in order to generate billing information.

11.6 Service Logic

11.6.1 Distribution

SSP

Initial Query Message Contents. SSP INVOKE (Provide Instructions): The SSP requests instructions for a call by sending a query message to the BSDB for processing. The query message contains the following parameters and data elements:[25]

Parameters	Data Elements
Package Type Identifier	Query with Permission
Originating Transaction ID	SSP assigned
Component Type	INVOKE (Last)
Invoke ID	SSP assigned :Operation Code Provide instructions. Reply required
Service Key	Automatic Number Identification (ANI) Calling number
Digits	Originating LATA
Digits	Originating station DN
Digits	Dialed (called number)
Digits	PIN (Private)
Digits	Traveling class mark
Originating Station Type	PVN or identified line. No special treatment
Automatic Call Gap (ACG) Encountered	Automatic Call Gapping

The service key identifies the service subscriber and the call processing records (CPRs) that should be used to obtain routing information and other details to be provided in the corresponding response message. If the digits length field of the dialed number or PIN has a value of zero, the SCP should send a response with routing instructions to the SSP indicating that the call should be routed to the PVN customer's attendant or a specific announcement. If a traveling class mark is included in the message, it should be used in conjunction with the service key to identify the customer and appropriate CPRs. If the BSDB fails to find the CPRs, a response message should be formatted that directs the appropriate network node to provide a vacant code announcement (VCA) or unassigned number announcement.

Additional Query Message Contents. SSP INVOKE (Provide Additional Instructions): The SSP requests additional instructions for routing a call by sending an additional query message to the BSDB for processing. This message is sent because the SSP encountered a return in the call treatment indicator of a trunk group parameter.

---

[25] BSDB: An SCP Application designed to support PVN service, Bellcore TA-TSY-000460.

SCP

PVN transaction types
1.   Standard PVN
2.   Standard PVN with resource counters
3.   PVN using PIN without resource counters
4.   PVN using PIN with resource counters.

Only the first transaction type, standard PVN is described in detail:

Read main temporary storage from PVN application queue

The PVN input from the SSP is placed on the PVN application queue by the data communications part of the SCP software. The first step in the PVN application processing in the SCP is to read the input from the SSP.

Interpret PVN TCAP input

The PVN application breaks down the TCAP components and interprets each one. In this example, there is a single component which forms a standard PVN transaction.

Extract station CPR key (ANI)

The station (service user) CPR is the call processing record that is associated with the line on which the PVN call originates. The calling number ANI is part of the input and forms the key for the station CPR.

Read station CPR

The PVN application issues a read request using the above mentioned key. This is a command to the data manager portion of the SCP software which does the PVN read processing.

Extract administrative CPR key (customer Number)

The station CPR contains the customer number which forms the key for the customer's administrative CPR.

Read administrative CPR

The PVN application issues a read request using the customer number key. This is a command to the data manager portion of the SCP software which does processing for the PVN application.

Ascertain which feature CPR(S) are needed

Having examined the information in the CPRs that have been read, the PVN application decides which (if any) feature CPRs are needed to process the PVN transactions.

Read feature CPR(s)

If feature CPRs are required to process the PVN call, the key(s) for the records are derived from data in the previously accessed station and administrative CPRs.

Origin screening processing

The PVN application checks the origin of the call to ensure that the caller is allowed to access the PVN network.

Call routing processing

The information required to route the call over the network (the response variable) is derived using data from the input (call variables) and data from the CPRs (intermediate variables).

Format TCAP response - routing information

The PVN application formats the response variables into a TCAP response component.

Write main temporary storage to output queue

Finally, the PVN application places the TCAP response on the output queue to be picked up by the data communications portion of the SCP software and sent back to the SSP which sent the original TCAP input.

BSDB Description

Note: Portions of the BSDB description were extracted from *BSDB: An SCP Application Designed to Support PVN Service*, Bellcore TA-TS-000460.

The Business Services Database (BSDB) is a generic SCP database application used to implement services like PVN. Service features for individual customers are encoded as CPRs in the BSDB. When a switching system with additional SSP capabilities recognizes a call requiring processing by a service implemented with BSBD, it queries the BSBD, interprets the message and the nodes contained in the CPRs, and the BSDB responds with instructions for handling the call.

For reliability, the BSDB should be deployed in pairs. Each BSDB in a mated pair contains the same set of CPRs, but the STP translations route calls with certain originating addresses to one BSDB and calls with other originating addresses to its mate. If one BSDB fails, the STPs redirect traffic to the other BSDB. The data in BSDBs at the SCPs are administered by the Service Management System (SMS). To handle a call, the BSDB should, via CPR instructions:

- Provide a domestic or international routing number
- Perform appropriate translations for network number calls
- Provide originating and terminating screening
- Instruct the SSP/Intelligent Peripheral (IP) to play announcements and collect digits
- Accept customer's authorization codes
- Provide for remote access to a customer's service
- Enable a customer to select interexchange carriers/international exchange carriers (ICs/INCs)
- Interface with private networks
- Provide a hop-off location
- Supply billing information for SSP recording
- Determine which resource counter is to be used (if appropriate)
- Specify the terminating treatment for the call
- Specify the originating treatment for the call.

The BSDB should be able to gather statistics on the calls it processes. There are three different types of call processing records:

- Administrative
- Station
- Feature.

Triggers

Any of the following can trigger PVN processing:

- Prefix or SAC
- ANI
- Called number from a remote access point.

11.6.2 Functional Flow

SSP - SCP

The following TCAP messages are required to support PVN for each of the PVN transaction types on page 219.

TCAP	Family name	Specifier	PVN transaction type
Invoke	Provide Instructions	Start	1,2,3,4
Invoke	Provide Instructions	Additional	1,2,3,4
Invoke	Connection Control	Connect	1,2,3,4
Invoke	Network Management	Auto. Call Gap	1,2,3,4
Invoke	Send Notification	Termination	2,3
Return Result	--	--	3,4
Invoke	Caller Interaction	Play Announ.	3,4
Return Error	--	--	1,2,3,4
Invoke	Procedural	Report Error	1,2,3,4
Reject	--	--	1,2,3,4

The following sample PVN logic flow represents the type of process which occurs in an SCP application host processor. Three of the transaction types are identified below, and shown in figures 38, 39, and 40.

1. Standard PVN transaction
   reads CPR records, develops routing information, and returns a response.
2. Standard PVN transaction with resource counter
   has resource counter processing. This type of transaction has two parts. The first input is the original query. The second input is termination information from the SSP, telling the SCP that the PVN call has ended, so that the resource counters can be updated.
3. Remote access PVN transaction using PIN
   The user is required to enter a personal identification code (PIN) to complete the call. This is also a two-part transaction. The first input is to gain access to the network. The second input is in response to an announcement which requests the user to provide the PIN and the number to be called.
   a. PVN using PIN without resource counters
   b. PVN using PIN with resource counters.

SCP - SMS

Messages flow between the SCP and the SMS to provide such functions as:

- Record updates
- Administration of usage data and statistics
- Status and exception reporting.

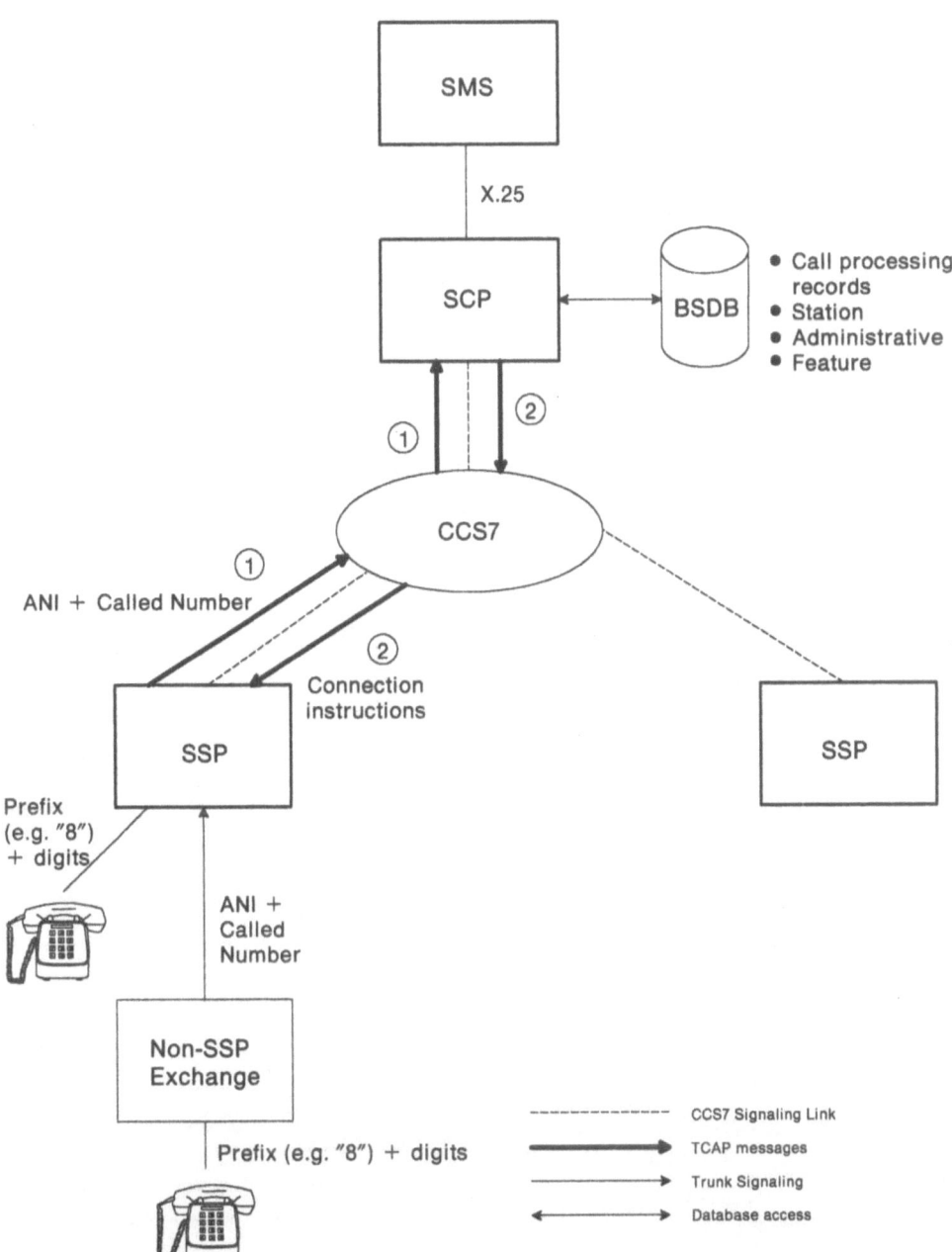

Figure 38. Standard PVN Transaction, Type 1

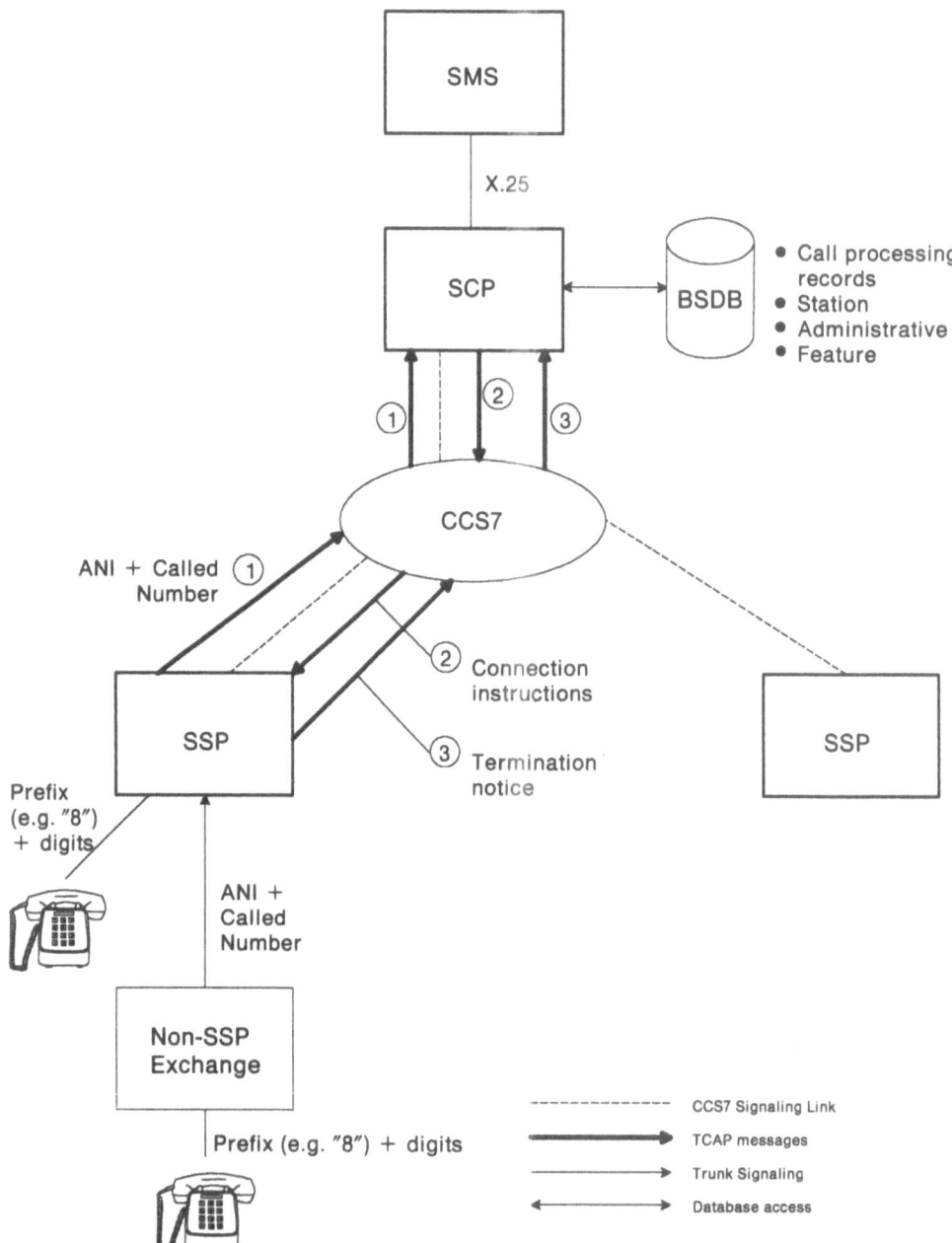

Figure 39. Standard PVN with Resource Counters, Type 2

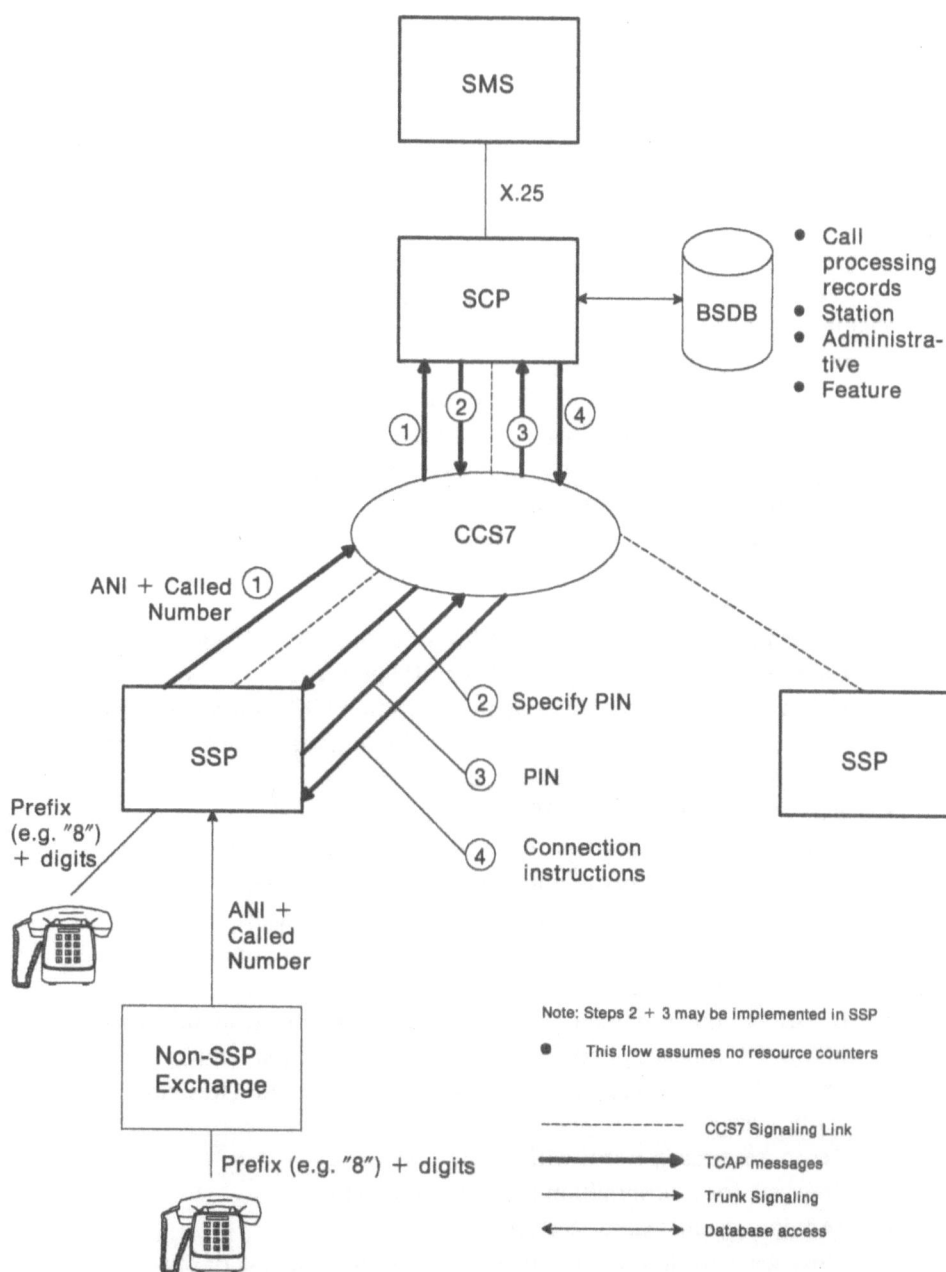

Figure 40. PVN with Remote Access Using PIN, Type 3

11.7 Traffic Measurement Requirements

SSP

Generic requirements are described in Chapter 2, "Functional Characteristics Common to Selected IN Services".

SCP

Note: The SCP description is extracted from *BSDB: An Application Designed to Support PVN Service*, Bellcore TA-TSY-000460.

1. The BSDB should maintain the following measurements (peg counts) on CCS7 messages and query processing:
    *   Query messages requesting BSDB service
    *   Query messages requesting additional instructions from the BSDB
    *   Query messages not matching the Originating Station CPR
    *   Incoming messages dropped because of an overload condition
    *   Query messages with error in data
    *   Query messages not responded to because the BSDB timer expired for a BSDB response
    *   Caller interaction messages sent for an authorization code
    *   Caller interaction messages sent for PIN and dialed number
    *   Call not allowed because Originating Station CPR not located (misrouted)
    *   VCA response messages sent (no Terminating Station CPR match)
    *   VCA responses sent because of CPR error
    *   VCA responses sent by CPR-specified action
    *   Changed Number Announcement response messages sent
    *   Disconnected number announcement response messages sent
    *   Unassigned number announcement response messages sent
    *   Resource Counter overflow announcement response messages sent
    *   Call not allowed announcement response messages sent
    *   BSDB TA timer expired for a response to the SSP (respond message)
    *   BSDB TA timer expired for a response from the SSP (caller interaction)
    *   BSDB TA timer expired for a response from the SSP (Resource Counter verification)
    *   Response messages sent with routing instructions
    *   Requests for a termination message to be sent
    *   Termination messages received
    *   Messages with termination indicators rejected because of overload
    *   Error messages received with termination indicators
    *   Error messages received with termination indicators caused by caller abandon or SSP failure
    *   Error messages received without termination indicators
    *   Error messages without termination indicators dropped because of overload

These measurements are taken in 30-minute intervals. At the end of each interval, the measurements are passed to the SCP node for storage with other measurements the node maintains on the BSDB.

2. The BSDB should maintain customer measurements (peg counts) about the number of valid and invalid PINs and authorization codes entered. These measurements should be maintained every hour, and at the end of each interval, the counters should be reset to zero.

3. The BSDB should send Exception Reports to the SMS when the following events occur:
   - CPR errors
   - Service Key in query message does not match a Station CPR
   - Sample storage exhaust
   - Resource Counter values are changed
   - Sampling is stopped because of a SCP overload
   - Sampling is resumed after a SCP overload
   - Customer level peg counts for valid and invalid authorization codes reach a prespecified maximum, as indicated in the Administrative CPR, within a 60-minute interval
   - Customer level peg counts for valid and invalid PINs reach a prespecified maximum, as indicated in the Administrative CPR, within 60-minute interval
   - Network management condition.

The BSDB should contain thresholds for reporting exceptions, maintained by the SMS. The threshold for this report may be set to allow transmission of only the first N reports in a five-minute period [where N is a variable determined by individual requirements].

The BSDB and SMS exchange three classes of administrative data; traffic and performance measurements, customer and network sample data, and control data.

Measurements: Every 30 minutes, the application passes its traffic and performance measurements to the SCP node. If the application fails or reinitializes, measurements for the current 30-minute period are lost. Measurement collection is unaffected if the BSDB is overloaded.

The SCP node combines the BSDB measurements with other common measurements it has taken on the BSDB in the same time interval. The SMS obtains these measurements by polling the SCP node.

11.8 Dynamic Requirements and Performance

SSP

The PVN user program should begin timing for a response from the SCP after sending a query message. A nominal value of three seconds should be used. If a response from the SCP is not received in this period, a reorder tone or an announcement should be given to the calling service user and the TCAP resources assigned when the query was sent should be released.

SCP

Note: The SCP description is extracted from "BSDB: An Application Designed to Support PVN Service", Bellcore TA-TSY-000460

The administrative response time for the BSDB is defined as the interval beginning from the receipt of an SMS command to the transmission of a BSDB response. The average response time should be from four to eight seconds, depending on the type and size of the SMS command message.

Administrative Capacity: The BSDB should be able to meet the following minimum administrative requirements:
- 20 updates or retrievals, per CPR per year
- 10 simultaneous special studies per customer
- Five network management control updates per hour
- 1.5 million customer-sampled calls per day (based on 10% sampling)
- 750,000 network-sampled calls per day (based on 5% sampling).

Number of Customers: Initially the BSDB should support at least 50 customers with 270,000 CPRs (25% small, 50% medium, and 25% large customers). The number of customers is expected to increase by 20% every year.

Resource Counters: The maximum number of Resource Counters that may be used per customer is 512.

SCP Response Time: Network response time for the BSDB is the interval beginning at the receipt of a query message and ending at the transmission of a response message.

During normal conditions, the average response time should be one second, and for 99% of all messages should not exceed 1.5 seconds. For the 10-second period after a BSDB first assumes the load of its mate, the mean response time should be 1.5 seconds, and should not exceed two seconds for 99% of all messages.

Overload Handling

Automatic Network Management Controls for SCP Overloads: Other services, like GNS and ABS have SCP overload controls. However, the BSDB overload control is based on controlling traffic from originating (calling) numbers or codes, while services such as GNS control destination (called) numbers. This is largely because queries are routed to the BSDB database on an originating basis, while queries for other services are routed to other databases on a destination basis.

Grade of Service Requirements

Single Line and Multiline Key Telephone Sets: The service delivery point-to-service delivery point grade of service blockage must not exceed a national average network busy hour blockage of 7% in the busiest month.

PBX: PVN must comply with the same grade of service requirements as those specified above.

Centrex: PVN shall comply with the same grade of service requirements as those specified above.

Facsimile Apparatus: PVN shall comply with the same grade of service requirements as those specified above.

Switched Data Service: PVN service delivery point-to-service delivery point grade of service shall provide a network busy hour blockage of no more than 7%.

Service Delivery Point: The service delivery point-to-service delivery point grade of service blockage shall not exceed a national average network busy hour blockage of 1% in the busiest month.

Chapter 12. PVN Application Description

The description in this chapter is based on IBM's and Siemens' views of

- Functional requirements and allocation. Requirements are specified and each item is assigned to the SCP, and the SSP, depending on where it must be developed.
- Functional units. Processing functions needed to handle the requested tasks are described and grouped logically into functional units for the separate components.
- Administrative Units. Processing functions not directly required in the real time call handling process are combined into administrative units.

A bold printed SSP means that the associated function must be performed in the Service Switching Point.

A bold printed SCP means that the associated function must be performed in the Service Control Point.

12.1 Functional Requirements and Allocation

Real Time Call Handling for Switched Voice Service
The functions required for this kind of PVN service are:

1. Announcements   **SSP**

    PVN must provide network intercept or recorded announcements as inherent network capability when a call or call attempt cannot be completed. At a minimum, such announcements must be provided for the following conditions:

    - Number disconnected
    - Number reassigned
    - Partial dial
    - Incorrect number of digits dialed
    - Time out during dialing
    - Network congestion
    - Preemption of access due to the provision of service for high-priority authorized users
    - Denial of access to off-net and international calls

- Denial of access to features
- Other related conditions.

This feature allows authorized personnel to record message announcements within the network. The recording is assigned an on-net number and is accessible by all stations.

2. Attendants   **SSP**

Attendants should be available 24 hours to assist users with dialing difficulties. Attendants should also be able to:

- Set up conferences and connect all conferees, if automatic network audio conferencing is not possible
- Verify authorization codes given verbally from all locations. Once a code has been verified, the attendant proceeds to set up the call.

A PVN user can call a PVN attendant to report a stolen or lost authorization code, or to obtain a credit adjustment for completed call to a wrong number, interruptions of calls, or unsatisfactory transmission.

3. Authorization Codes   **SSP/SCP**

PVN authorization codes are used for:

- User identification
- Class of service
- Originating on and off-net calls if identification cannot be made by other means for billing and class of service purposes
- When override capabilities are desired.

The class of service derived from an authorization code takes precedence over that derived from other means.

While the network operator determines the dialing sequence for entering an authorization code, the dialing sequence should be such that the PVN network is alerted that an authorization code is about to be entered, so as not to delay the processing of calls not requiring the use of this feature.

Repeated attempts that are closely spaced in time from one or more locations using invalid authorization codes or other identifiable patterns, are detected by procedures implemented by the network operator.

4. Closed User Group (Call Screening)   **SSP/SCP**

Call screening comprises a set of features that determine a call's eligibility to be completed as dialed, based upon class of service information associated with the user, the station, or the trunk group.

5. Class of Service   **SSP/SCP**

The class of service is determined from the ANI, authorization code, traveling classmark, or trunk group. The class of service derived from an authorization code takes precedence over that derived from other means. Classes of service identify access and feature restrictions:

- Access restrictions include (but are not limited to) access to off-net calling, access to portions of the PVN, and access to other than specified numbers.
- Feature restrictions allow or restrict access to network features by users or groups of users.

Time of day, day of week, and day of year restrictions are possible, implemented on a per station, per location, and per authorization code basis. A class of Service Override feature allows an individual PVN user to override the terminal equipment's preassigned class of service treatment for the duration of the call by entering a valid authorization code. Call detail records reflect all relevant data on the call, which is charged to the authorization code user rather than to the originating station.

6. Traveling Classmark   **SCP/SSP**
This feature provides for the acceptance of traveling classmarks from any PVN locations served by PBXs and Centrex. User calling characteristics and limitations may be determined from the traveling classmark.This feature is made available to all types of representative terminal equipment.

7. Code Block   **SCP/SSP**
This feature screens ineligible users, stations, and trunks with certain class of service access restrictions and prevents them from calling specified area codes, exchange codes and countries by translating the dialed digits and comparing the results with the class of service. Blocked calls are directed to the network recorded announcements.

8. Network Audio Conferencing   **SCP/SSP**
Conference call support is provided for a mixture of on-net, off-net, and virtual on-net stations, and includes the following conference features:
- "Meet-Me" - A number of users are connected by dialing an access code at a predetermined time or as directed by the attendant.
- Present - Any number of authorized users place such calls, with the PVN allowing call lists of a number of conferees.
- Add-on - The call controller can add conferees to an existing conference.
- Attendant-assisted - used primarily by callers who cannot initiate a conference call from their telephones.

9. Special Access   **SSP/SCP**
Inward station access allows off-net callers to be connected to predesignated PVN stations by dialing certain customer GNS or other GNS numbers. Inward Selected Access shall allow off-net callers to be rerouted to network locations by dialing specified GNS numbers. Once the caller has been connected, the network provides a recorded announcement identifying the DTMF selection digit(s) that provide further routing of the call.

Real Time Call Handling for Switched Data Service
The functions required for this kind of PVN service are:

1. PVN Access   **SSP/SCP**
    Dedicated PVN lines and trunks
    The network operator should be able to assign certain lines or trunks as dedicated PVN lines or trunks. All calls originating on such lines or trunks should be treated as PVN calls, without the need for identification.

Business group lines

All calls dialed by a business group line with a prefix assigned to indicate a PVN call for that business group customer should be treated as PVN calls. Provision should also be made for calls using Automatic Flexible Routing (AFR).

Non-dedicated PVN lines

It is planned to assign a service code (SC) that may be used on certain lines (such as PBX lines) to access a PVN. Translations should be provided to indicate if this code should be accepted from a line.

Direct connect line

The network operator should be able to assign certain lines within a business group as PVN direct connect lines. All originations on such lines should be treated as PVN calls.

The above capabilities should be provided so that they are not limited to PVN calls. That is, a trigger table should be provided to identify calls from lines or trunks with certain classes of service or dialed with certain prefixes as calls requiring database queries.

A member of a PVN may originate a PVN call while connected to another party (for example, using the add-on or transfer features). When this occurs, the SSP should make a regular PVN query to the SCP and use the information in the response message to establish the third leg of the call.

This feature may require downloading of triggers from the SCP to the SSP.

2. PVN dialing plan

After a customer has indicated that he is placing a PVN call as described, he can complete the call by using a predetermined dialing plan. If the digits received on a PVN call do not agree with the dialing procedure, the call is blocked at the PVN switch and reorder announcement returned.   **SSP**

Incoming business group tie-trunks may carry PVN calls. The network operator can specify for each incoming tie-trunk whether all calls on the tie-trunk should be treated as PVN calls or if the calls received should be treated the same as calls originating from stations within the business group. Calls over certain tie-trunks may include a private network traveling classmark. The network operator should be able to specify incoming PVN access tie-trunks on which the final digit(s) are defined is a private network traveling classmark. If a traveling classmark is received, it should be included in the query message to the SCP. The switch can receive rotary dial, DTMF, and MF signaling on incoming PVN tie-trunks.   **SSP/SCP**

A "station group designator" may be received on a direct trunk from a PBX. Station group designators are identical to the traveling classmarks described above, and should be treated in the same way.   **SSP/SCP**

3. Remote access dialing plan

A member of a PVN should be able to place a call from any remote station to access the PVN. Provision should be made for the network operator to assign a number, or numbers, to a PVN customer for this purpose. When an SSP receives a remote access number, the call should be handled in one of the following ways:

- The network operator should indicate that calls to certain directory numbers result in a query being made to an SCP database. Translations indicate (for each of these directory numbers) the ANI information digits and the subsystem numbers to be included in the query message. With this capability, the network operator can provide other data base services which can be reached by dialing of a directory number.

  For PVN service, the SCP would then request that an announcement or recall dial tone be connected and that a specific number of digits be collected from the caller as a PIN. (A PIN is used to identify the caller or his capabilities). The message from the SCP would also specify that a second announcement or recall dial tone be connected and the normal PVN dialing sequence for the PVN customer be collected. A second message should then be sent from the SSP to the SCP reporting the digits received. The SCP responds to this message with routing and billing information for the call.   **SSP/SCP**

- A particular switch's remote access capability could be modified to provide PVN remote access so that on receipt of a PVN remote access number, the SSP (without making a database query) returns an announcement or recall dial tone to the caller who enters a PIN. The network operator should be able to specify for each PVN customer the number of digits to be collected as a PIN. Having received a PIN of the proper length, the SSP should connect another announcement or recall dial tone to the caller. The caller should then be able to dial a PVN call as though calling from a station that was part of the PVN. A query is made to the SCP to determine if the PIN is correct, and to obtain routing and billing information for the call.   **SSP**

If the caller enters an incorrect PIN, the SCP sends another message to the SSP requesting it to connect an announcement or recall dial tone, and to collect a specific number of digits as a PIN. Having received these digits from the caller, the SSP should send another message to the SCP transmitting the digits received. The number of times a caller may dial an incorrect PIN without being disconnected is included in the PVN customer's data in the SCP. When this number is exceeded, the SCP instructs the SSP to abort the call.   **SSP/SCP**

The SSP should be arranged to accept the DTMF signal "*" while receiving either the PIN or the normal PVN dialing sequence on a remote access call. The DTMF signal "*" is used by the caller to indicate that he has made a mistake. When the "*" signal is received, the SSP should discard the digits already received and be prepared to receive the entire sequence of digits.   **SSP**

The SSP should return answer supervision to the originating office when connecting the first announcement or the first recall dial tone to the caller. At least one cycle of audible ringing should be returned to the caller before the start of the announcement or the recall dial tone.   **SSP**

4. PVN calls from PBXs using multifrequency signaling

    An SSP should be able to receive calls using MF signaling from PBXs so equipped. The switch should be able to accept normal signaling on these trunk groups.   **SSP**

5. SCP Queries   **SSP/SCP**

    The SSP sends a query to the SCP under the following conditions:

    a. Dialing has been completed on a PVN call from a station or tie line served by the SSP

    b. A PVN call has been received:

        *   From a non-SSP switch

        *   Via remote access

        *   Via remote access and the caller has completed dialing PIN and the PVN dialing sequence.

    c. An extension call requiring database routing information has been received.

6. Completed as requested   **SSP/SCP**

    The actions taken on a given call depend on the response message received from the SCP. This response should include information needed to route the call and may request the SSP to send additional information, such as whether the call was answered and, if so, the call duration.

    When the SCP requests the SSP to send termination information, the response message from the SCP should contain the transaction ID assigned by the SSP. The response should contain an INVOKE component which instructs the SSP to route the call based on the value of the data elements in the message. The following actions may be requested in the response message:

    *   Route the call over the public network

    *   Connect an announcement to the calling station

    *   Route the call over private facilities

    If an AMA record is made, a billing indicator field is included in the response message.

7. Authorization codes

    Instead of sending a response as a result of a query message, the SCP may request that an authorization code be obtained from the caller. This is done by sending a "Conversation with Permission" message with an INVOKE "Play Announcement and Collect Digits".   **SCP**

    This feature should be provided so that its use is not restricted to PVN service. That is, for any service for which the SSP has sent a query and is waiting for a response, the SSP should be able to accept an INVOKE "Play Announcement and Collect Digits", and forward the digits collected to the SCP.   **SSP**

    Having received this message, the SSP should make the necessary connections to play the requested announcement to the caller and to accept DTMF digits from the customer. This capability is normally provided by means of an optional separate intelligent peripheral, but the SSP should also be able to play announcements and collect digits without using the intelligent peripheral. If the calling station is connected directly to the SSP, the SSP should also be prepared to accept rotary dial digits from the caller. The number of digits to be collected is specified in a "Number of Digits" field.   **SSP**

    Receipt of the first digit from the caller removes the announcement. If the switch cannot detect a digit while playing an announcement, it should play the announcement once, connect recall dial tone, and be prepared to detect digits. ("At the tone, please ...")   **SSP**

In many cases, the announcement specified in the message from the SCP is "recall dial tone", in which case, a recall dial tone should be connected in the normal manner.   **SSP**

Having received the authorization code from the caller, the SSP should send a RETURN RESULT message to the SCP. The message should indicate "Caller Interaction" and include the digits entered by the caller. The SSP should also restart the T1 timer that was used when the original query message was sent to the SCP.   **SSP/SCP**

If the caller remains off-hook for five seconds and dials no digits, the SSP responds as described in the previous paragraph. However, the "Digits Length" field has a value of zero (indicating that no digits were received) so that the SCP can return routing instructions to the SSP to route the call to the PVN customer's attendant. The attendant could then set up a new call and enter all information required to complete the call.   **SSP/SCP**

If the caller remains off-hook and dials insufficient digits, but at least one digit, the SSP sends a RETURN RESULT. This RETURN RESULT includes the digits entered by the customer, so that the SCP can ask the customer to try again or to send instructions to terminate the call. Normal partial dial timing should be used to detect this condition.   **SSP/SCP**

If the caller abandons the call without dialing an authorization code, the SSP should send a Response message to the SCP with a RETURN RESULT including the Standard User Error Code "Caller Abandon".   **SSP/SCP**

The SSP should be arranged to accept the DTMF signal "*" while receiving an authorization code (like during receipt of PIN or normal call). When the "*" signal is received, the SSP should discard the digits already received and be prepared to receive the entire sequence.   **SSP**

If there is a protocol error in the message from the SCP, the SSP should respond with an error message with "REJECT" as its Component Type Identifier. For other errors in the SCP message, the error message sent by the SSP should indicate "RETURN ERROR". The SSP should also connect reorder tone or reorder announcement to the calling party when an error condition occurs.   **SSP**

On calls originating in non-SSP end offices, the SSP should return answer supervision to the originating office before connecting the announcement. If CCS7 signaling is used for call setup from the originating office, and a subsequent Address Complete Message and/or Answer Message is received from the far end, an Address Complete Message and/or a second Answer Message should not be sent to the originating office. If, instead of an answer, a Release Message with Cause is received, the SSP should connect the tone or announcement required for the particular cause.   **SSP**

8. Remote access calls

If the arrangement described in " Remote Access Dialing Plan " is provided for Remote Access, the SCP does not send a response to the query message, but requests that additional information be obtained from the caller. The "Conversation With Permission" message contains two INVOKES, each with the Operation Code "Play Announcement and Collect Digits". The first

INVOKE specifies the number of digits to be entered as a PIN, the second indicates that the SSP should determine the number of digits to collect using the normal translation information for the particular PVN customer.   **SSP/SCP**
If the switch cannot detect a digit while playing an announcement, it should play the announcement once, and then connect recall dial tone and be prepared to detect digits. Having received both sets of digits from the caller, the SSP sends two RETURN RESULT components in a "Conversation with Permission" message to the SCP. Each RETURN RESULT should indicate "Caller Interaction" and include the numbers received from the caller.   **SSP**
On calls originating in non-SSP end offices, the SSP should return answer supervision to the end office before connecting the announcement. If CCS7 signaling is used for call setup from the end office, the SSP should subsequently connect a tone or announcement on the call if a Release Message with Cause is received from the far end. If an Address Complete Message and/or an Answer Message is received from the far end, the SSP should not repeat them to the originating office.   **SSP**
If the caller remains off-hook for five seconds and dials no digits, the SSP sends a single RETURN RESULT component to the SCP. The "Digits Length" field has a value of zero (indicating that no digits were received) so that the SCP can instruct the SSP to route the call to an attendant to assist the caller. Similarly, if no digits are received from the caller by a switch that normally collects a PIN and a dialed number prior to sending a query, the switch should send a query with the "Digits Length" part of the PIN digits field having a value of zero.   **SSP/SCP**
If the caller has entered an incorrect PIN with either of the Remote Access call procedures, the SCP may send a message requesting the SSP to have the caller redial their PIN. This is another "Conversation With Permission" message containing a single INVOKE to "Play Announcement and Collect Digits". The SSP should connect the announcement and collect the number of digits requested. Having received the digits, the SSP sends a RETURN RESULT in a "Conversation With Permission" message to the SCP. This RETURN RESULT indicates "Caller Interaction" and includes the numbers received from the caller in a digits field. If the PIN is still incorrect, the SCP sends a response message requesting that the SSP either collect the digits again, or connect a reorder tone or reorder announcement.   **SSP/SCP**
The SSP should be arranged to accept the DTMF signal "*" while receiving either the PIN or the normal PVN dialing sequence on a remote access call. The DTMF signal "*" is used by the caller to indicate that they have made a mistake. When the "*" signal is received, the SSP should discard the digits already received and prepare to receive the entire sequence of digits.   **SSP**

9. Resource counters
The SCP will probably contain resource counters for routes between switch locations for many PVN customers. When the SCP receives a query message for a PVN call, it checks the resource counter associated with the particular route. If the resource counter indicates that there is an available resource, it increments the counter and sends a response message to the SSP instructing it

to complete the call. The response message also requests the SSP to send termination information at the end of the call so that the associated resource counter can be decremented.   **SSP/SCP**

The response message carrying routing instructions or the termination message for a call involving a resource counter may get lost, therefore, means must be provided to verify that the resource counter values are correct. To do this, the SCP periodically sends queries to the two SSPs associated with a particular resource. This query requests the SSPs to check the number of existing calls using the particular resource and reports this number back to the SCP. The SCP then adds the numbers received from the two SSPs, and corrects the resource counter if necessary. This procedure normally takes place during periods of low traffic.   **SSP/SCP**

10. Terminating calls   **SSP/SCP**

Calls terminating at PVN Stations

The network operator should be able to indicate (in the switch translations for both SSP switches and non-SSP switches) PVN stations that are restricted from receiving non-PVN calls. The only calls to be completed on these lines are those tagged with a PVN indicator.

Hop-off Calls and Calls to Private Facilities

Existing Business Group capabilities permit a call received over a private facility to hop-off using AFR lists or to be connected to other private facilities. These same capabilities should be provided for PVN calls received over public facilities when CCS7 signaling is used for call setup between the originating SSP and the terminating PVN switch.

The following is a list of the TCAP messages required for PVN:

- INVOKE with Provide Instructions
- INVOKE with Connection Control
- INVOKE for Network Management
- INVOKE with Send Notification
- RETURN RESULT
- INVOKE with Caller Interaction
- RETURN ERROR
- INVOKE Procedural (Report Error)
- REJECT.

SMS-SCP Transactions

1. The SCP and the SMS must exchange messages to update the BSDB in the SCP.   **SCP**

2. Two classes of message are defined in the SCP/SMS interface: measurement data and status data.   **SCP**

   - Measurement data
     The SMS collects data from the SCPs about SCP performance and traffic characteristics. The SCP software must be designed to provide the SMS with separate sets of measurements for the SCP operating system, the SCP support software, the signaling traffic between the Signal Transfer Points (STPs) and the SCPs and each application implemented at the node.

Measurement messages record hardware failures, software failures and the volumes of traffic in and out of the SCP. The SCP node transmits the set of measurements that are collected by the applications and then stored at the SCP node for later transmission to the SMS.

Periodically, the SMS requests each SCP node to transmit one interval of data collected since the last request. The SMS validates the data and stores it for administrative use.

- Status data

  The SMS should be informed of SCP performance and should take action to minimize SCP performance degradation caused by traffic overload, to minimize data loss when either the SMS or the SCP fails and to maintain the consistency of the SMS and SCP databases.

  The status messages from the SCP to SMS report on application availability, SCP overload status and the assumption and later release of mate traffic by an application.

  The Transport Service and Application Service protocols provide essential features, some of which are used by SCP applications and other network elements that communicate with operations systems.

3. The BSDB and SMS exchange three classes of administrative data: **SCP**
    - Traffic and performance measurements
    - Customer and network sample data
    - Control data.

Traffic Measurements

1. Separate counts should be made for irregularities occurring during SSP PVN calls.   **SSP**

2. Base count measurements should be made so that the volume of PVN calls can be monitored. The SSP should count all PVN calls originating in the SSP switch that have reached the dialing complete stage and all PVN calls that have been received from a non-SSP switch.   **SSP**

3. The BSDB should maintain peg count measurements on CCS7 messages and query processing. The measurements should be taken in 30-minute intervals. At the end of each interval, the measurements (and the date and time at which the interval ended) should be passed to the SCP node for storage with other measurements the node maintains on the BSDB.   **SCP**

4. The BSDB should maintain measurements (peg counts) on a customer level. These measurements should be maintained in 60-minute intervals. At the end of each interval, the counters should be reset to zero.   **SCP**

5. The BSDB should send Exception Reports to the SMS on the occurrence of irregularities.   **SCP**

6. The BSDB should contain exception threshold data (maintained by the SMS) for controlling the generation of Exception Reports. The threshold for this report may be set to allow transmission of only the first N reports in a five-minute period.   **SCP**

7. Every 30 minutes, the BSDB should pass its traffic and performance measurements to the SCP node. If the BSDB fails or reinitializes, measurements

for the current 30-minute period are lost. If the BSDB enters an overload state, measurement collection is unaffected.   **SCP**

8. The SCP node combines the BSDB measurements with other common measurements it has taken on the BSDB in the same time interval. The SMS accesses these measurements by polling the SCP node.   **SCP**

Billing Measurements and Statistics

1. The network operator must provide a system for collecting, aggregating and formatting all billing information for PVN services. Call details must be identified to the lowest level of call origination possible: the originating station's calling number, location number or authorization code.   **SSP**

2. Station identification may be provided by Automatic Identification of Outward Dialing (AIOD), Automatic Number Identification (ANI),or an authorization code. The CCS7 station identification may also be used within PVN for call screening purposes. Switched digital integrated service for ISDN provides station identification as specified in CCITT Recommendation Q.931. Call details include the calling number, the date of message, and the time of message origination.   **SSP**

3. An AMA record should be made on all originating PVN calls on which Billing Indicator information is included in the response message from the SCP. The call code used is included in the billing indicators information. Additional call codes will probably be assigned for services provided by an SSP. Therefore, the SSP should be able to record any call code received from the SCP.   **SSP**

Exceptions

1. Automatic network management controls for SCP overloads. When the calling rate from a number of call origination areas becomes excessive, the SCP can produce an overload condition. To protect the SCP from the effects of overloads, an automatic node manager control (called the SCP overload control) is required. The SCP overload control should operate so that the SCP node detects when it is overloaded by measuring the delay between receiving a query and returning a response. When the SCP node determines that it is overloaded, it should notify its applications (such as BSDB) with a message. The applications should respond with certain actions to help alleviate the overload. The actions taken depend on the level of overload.   **SCP**

2. Other services are also equipped with SCP overload controls. There is, however, a fundamental difference between BSDB and other services in the activation of an overload control. BSDB control is based on controlling traffic from originating (calling) numbers or codes, while services such as GNS control destination (called) numbers. The primary reason for this is that routing of queries to the database for BSBD service is done on an originating basis, while routing of queries to the database for the other service is normally done on a destination basis. (Destination in this section is used in a broad sense; for GNS, the destination could be a POTS number or an GNS number, for ABS, destination could be the calling card number in the database, which is usually different from the originating station number). Because the objective is to

protect the database from overload, control of attempts to the database is based on how these attempts are routed to the database.   **SCP**
3. If the SCP node enters an overload condition, it passes a message to the BSDB application that is overloaded, and the message contains the overload level. In response, the BSDB should take steps to reduce both the administrative activity and call processing loads. The SCP node should also send a message when it exits from an overload level. The message gives the new overload level, which includes level 0 (no overload).   **SCP**

12.2 Functional Units in SSP

Initial Call Handling
This unit is responsible for initial PVN call handling.

Processing Functions
1. The unit decides whether a call is a PVN call or not (this may be done by the SSP number translator).
2. After a caller has indicated that he is placing a PVN call, he should be able to complete the call by using a predetermined dialing procedure.
3. If the digits received on a PVN call do not agree with the dialing procedure, the call should be blocked, and a reorder announcement returned.
4. PVN calls usually query the SCP to obtain the calling characteristics.
5. PVN dedicated access:
   - Lines.
   - Trunks.
   - Trunk groups.
   - Tie-trunks.
   - Business group lines with dialed PVN prefix.
   - Business group lines assigned as direct connect lines.
   - Tie-trunks with travelling classmark. (Access to an SCP might be necessary).
   - Direct trunks from PBXs.
   - Direct trunks from PBXs with station group designator.
6. PVN non-dedicated access:
   - Non-dedicated access is initiated by dialing a special prefix, Special Code (SC), or Special Access Code (SAC).
   - Identification is made using caller data:
     – Authorization code or PIN from individuals
     – Travelling classmark from individuals, PBXs or Centrex.
   - The initial call handling unit calls the transaction handling unit to send a TCAP request to the SCP. The SCP controls the collection of the identification data by sending an answer request to the SSP.

7. PVN remote access
   - Remote access calls are recognized as calls to certain specified directory (network) numbers, or "remote access numbers", that is PVN specific private numbers.
   - With specified directory numbers, the initial call handling unit calls the unit transaction handling to send a TCAP request to the SCP and to handle the further interaction unit with the caller and the SCP.
   - With "remote access numbers", the SSP controls the collection of the identification data (PIN) without assistance from the SCP.

Transaction Handling for PVN Access
This unit controls different transaction types.

Processing Functions
1. There are three transaction types:
   - **Standard PVN transaction** reads CPR records, develops routing information, and returns a response.
   - **Standard PVN transaction with resource counter.** This is the same as the standard PVN transaction, with the addition of resource counter processing. This transaction has two parts: the original query and termination information (that tells the SCP that the PVN call has ended, so that the resource counters can be updated).
   - **Remote access PVN transaction using PIN.** The user is required to enter a personal identification code (PIN) to complete the call. This is also a two part transaction. The first part is to gain access to the network, the second is to respond to an announcement which requests the user to provide the PIN and the number to be called. This transaction type may be used with or without resource counters.
2. Only the first transaction type is described in the following:
   - Prepare the query message for the SCP, with ANI as station key.
   - Wait for a response.
   - Accept the response or call authorization code handling.
   - Route the call.
   - Process call termination using data from the response.
3. Authorization code handling:
   - Instead of responding to a query message, the SCP may request that an authorization code be obtained from the caller, by sending a "Conversation With Permission" message with an INVOKE "Play Announcement and Collect Digits".
   - This feature should be provided so that its use is not restricted to PVN service. With any service on which the SSP has sent a query and is waiting for a response, it should be able to accept an INVOKE "Play Announcement and Collect Digits" and to forward the digits collected to the SCP.
   - The SSP should play the requested announcement and accept DTMF digits from the customer (*Call Announcement Handling* unit).

- Having received the caller's authorization code, the SSP should send a RETURN RESULT message to the SCP,indicating "caller' interaction" and including the digits given by the caller. The SSP should also restart a timer used when the original query message was sent to the SCP.

4. If the caller is off-hook for five seconds and dials no digits, the SSP sends a response. However, the "Digits Length" field will have the value of zero, indicating that no digits were received, thus permitting the SCP to return routing instructions to the SSP, and to route the call to the PVN customer's attendant. The attendant can establish a new call and enter all the information required to complete the call.

5. If the caller is off-hook and dials insufficient digits (but at least one digit) the SSP should send a RETURN RESULT including the digits given by the customer. This way, the SCP can ask the customer to try again, or send instructions to abort the call. Partial dial timing should be used to detect this condition.

6. If the caller abandons without dialing an authorization code, the SSP sends a response message to the SCP with a RETURN RESULT including the standard user error code "Caller Abandon". If there is a protocol error in the message from the SCP, the SSP responds with an error message with "REJECT" as its component type identifier.

7. For other SCP message errors, the SSP error message sent should indicate "RETURN ERROR". The SSP should connect a reorder tone or reorder announcement to the calling party when an error occurs.

8. For calls originating in non-SSP end offices, the SSP should return answer supervision to the originating office before connecting the announcement. If CCS7 signaling is used for call setup from the originating office, and a "subsequent address complete" message and/or "answer" message is received from the far end, an "address complete" message and/or a second "answer" message should not be sent to the originating office. If a release message with exception is received, instead of an answer, the SSP should connect the tone or announcement required for the particular exception.

9. Repeated attempts within a short period of time using invalid authorization codes or other identifiable patterns are detected by the network operator.

Transaction Handling for Remote Access
Processing Functions
Having received a remote access number, this unit handles the call in one of the following ways:

1. With calls to defined directory numbers, a query is made to the SCP. Translations for each of these numbers should indicate the ANI information digits and the subsystem numbers to be included.

   The SCP does not send a response to the query message, but requests that additional information be obtained from the caller (in a manner similar to that described for authorization codes). The "Conversation With Permission" message contains two INVOKES with the operation code "Play Announcement and Collect Digits". The first INVOKE specifies the number of digits to be entered as a PIN, the second indicates that the SSP should determine the number of digits to collect using the normal translation information (dialing sequence) for the particular PVN customer.

After receiving both sets of digits from the caller, the SSP sends two RETURN RESULT components in a "Conversation With Permission" message to the SCP. Each RETURN RESULT indicates "caller interaction" and includes the caller's number.

For calls originating in non-SSP end offices, the SSP should return answer supervision to the end office before connecting the announcement. If CCS7 signaling is used for call setup from the end office and a "release message with cause" is received from the far end, the SSP connects a tone or announcement on the call. If an address complete message and/or an answer message is received from the far end, the SSP should not repeat them to the originating office.

If the caller is off-hook for five seconds and dials no digits, the SSP sends a single RETURN RESULT component to the SCP. The "digits length" field should have the value of zero, indicating that no digits were received. In this way, the SCP can instruct the SSP to route the call to the PVN customer's attendant.

If no digits are received from the caller by a switch that normally collects a PIN and a dialed number prior to sending a query, the switch should send a query with the "Digits Length" part of the PIN digits field having the value of zero.

If the caller enters an incorrect PIN during either of the remote access call procedures, the SCP may send a message requesting the SSP to have the caller redial his PIN. This requires another "Conversation With Permission" message containing a single INVOKE to "Play Announcement and Collect Digits". The SSP connects the announcement and collects the number of digits requested. After receiving the digits, the SSP sends a RETURN RESULT in a "Conversation With Permission" message to the SCP. This RETURN RESULT indicates "caller interaction" and includes the numbers received from the caller in a digits field. If the PIN is still incorrect, the SCP sends a response message requesting that the SSP collect the digits again, or connect the reorder tone or announcement.

2. If the switch has a remote access capability, this capability can be modified to provide PVN remote access.

On receipt of a PVN remote access number, the SSP (without making a database query) returns an announcement or recall dial tone to the caller, who enters a PIN.

On receipt of a PIN of the proper length, the SSP should connect another announcement or recall the dial tone.

The caller is then able to dial a PVN call, as though placing the call from a PVN station. A query is made to the SCP to determine whether the PIN is correct and to obtain routing and billing information for the call.

If the caller enters an incorrect PIN, the SCP may send another message to the SSP, asking it to connect an announcement, or recall a dial tone, and to collect a specific number of digits as a PIN. Having received these digits, the SSP sends another message to the SCP transmitting the digits received. The number of times that a caller may dial an incorrect PIN without being disconnected is included in the PVN customer's data in the SCP. When this number is exceeded, the SCP instructs the SSP to abort the call.

The SSP should return answer supervision towards the originating office when connecting the first announcement or the first recall dial tone to the caller. At least one cycle of audible ringing should be returned to the caller before the start of the announcement or the recall dial tone.

SCP Response Handling

Processing Functions

1. Processing calling characteristics
   - Calling characteristics (that is, access and feature restrictions) are defined for individuals, or equipment representing individuals, or groups of individuals.
   - Access to (or retrieval of) calling characteristics stored in the SCP is done:
     - Via ANI or PIN for individuals (traveling classmark and authorization code are not used as access keys)
     - Via ANI, address, identifying number, or designator of equipment such as lines, trunks, tie-trunks, trunk groups, PBXs, Centrex, terminals, stations.
   - An exception to this rule is the "algorithmic" code block feature requiring no memory.
   - The following are defined and processed (class of service):
     - Access restrictions concerning:
       - Off-net calling
       - Portions of PVN
       - Other than specified numbers
       - Area codes
       - Exchange codes
       - Country codes
       - Closed user group.
     - Class of service override only for individuals identified via an authorization code
     - Feature restrictions concerning use of network audio conferencing
     - Additional time of day, day of the week, and day of the year restrictions shall be possible.
   - A query to the SCP database is necessary if the SSP has no knowledge of the calling characteristics of the caller or equipment. Bellcore specifications indicate that the SSP will have no knowledge of the calling characteristics of individuals, but it might be advantageous to store calling characteristics related to equipment within the SSP, as the volume and probability of update is small. This requires the determination of calling characteristics and the update function to be implemented by the SSP.

2. Number translations; inward station access allows the connection of off-net callers to predesignated PVN stations by dialing GNS numbers. Inward selected access allows the rerouting of off-net callers to network locations by dialing specified GNS numbers. Once the caller has been connected, the network provides a recorded announcement, identifying the DTMF selection digits that provide further routing of the call.

Routing and Call Termination

Processing Functions

1. The actions to be taken depend on the response message received from the SCP. This message includes information needed to route the call and may request the SSP to send additional information (such as whether the call was answered and, if so, the call duration) to the SCP.
2. The SCP response message contains the transaction ID assigned by the SSP and an INVOKE component which instructs the SSP to route the call, based on the value of the data elements. The following actions may be requested in the response message:
    *   Route the call over the public network
    *   Connect an announcement to the calling station
    *   Route the call over private facilities.
3. If an AMA record is made, the response message includes a billing indicators field.
4. If the SCP contains resource counters for routes between switch locations, the response message requests the SSP to send termination information at the end of the call, so that the associated resource counter can be updated.

Network Audio Conferencing

Processing Functions

Conference call support is provided for a mixture of on-net and off-net stations and includes "Meet me", Present, Add-on, and Attendant-Assisted conferences.

Resource Counter Handling

Processing Functions

1. As the response message containing routing instructions or a termination message for a call involving a resource counter may be lost, a function must verify that the resource counter values are correct.
2. To do this, the SCP sends queries to the SSP's associated with a particular resource counter during periods of low traffic.
3. A load run for resource counters must also be possible (in batch mode) during SCP startup.

Announcement Handling

This unit plays an announcement and collects digits. (Announcements concerning error situations are not covered by this unit.) The PIN, authorization code, and normal PVN dialing sequence are collected. This unit reports the digits collected, and such errors as timeout (where the caller remains off-hook for five seconds without dialing), the caller remaining off-hook and dialing insufficient digits, or call abandonment without dialing.

Processing Functions

1. The SSP can play announcements and collect digits without using the intelligent peripheral. If the calling station is directly connected to the SSP, the SSP should also be prepared to accept rotary dial digits from the caller. The number of digits to be collected is specified in a "Number of Digits" field.
2. Receipt of the first digit from the caller removes the announcement. If the switch cannot detect a digit while playing an announcement, it should play the announcement once, and then connect recall dial tone and prepare to detect digits. ("At the tone, please ...")
3. The SSP should be arranged to accept the DTMF signal "*" while receiving a PIN, a normal PVN dialing sequence, or an authorization code.
4. The DTMF signal "*" is used by the caller to indicate that he has made a mistake. When the "*" signal is received, the SSP discards the digits already received, and prepares to receive the entire sequence.
5. The announcement specified in the message from the SCP is often "recall dial tone". In this case, recall dial tone should be connected in the normal manner.

Attendants

Attendants in a PVN must have certain capabilities.

Processing Functions

1. Attendants should be available 24 hours a day to help users with any dialing difficulties. They should also:
   - Be able to set up conferences and connect all conferees, if automatic network audio conferencing is not possible
   - Be able to verify authorization codes given verbally from all locations. Once a code has been verified, the attendant sets up the call.

2. A PVN user can call a PVN attendant to report a stolen or lost authorization code or to obtain a credit adjustment for a completed call to a wrong number, interruptions of calls, or unsatisfactory transmission.

AMA Generation
This unit is responsible for generating the appropriate AMA record.

Processing Functions
1. At call completion, the SSP produces an AMA record based on the billing indicator field from the SCP. AMA is usually handled in the CP.
2. The AMA record contains information about the calling party, called party, billing number, call duration, answer condition, and so on.
3. The AMA records are stored on disk and later transferred to a billing center that generates the actual customer bills.

Traffic Measurements
This unit maintains all required traffic and maintenance measurements relating to PVN.

Processing Functions
1. Peg counts and usage counts are provided with a recording period of thirty minutes. Most of these measurements apply to all services.
2. Separate counts should be made for irregularities occurring on SSP PVN calls.
3. Application measurements are done in 30-minute intervals, on the hour. At the end of the interval, the measurements are passed to the operations and support systems.

Automatic Call Gapping
This unit handles Automatic Call Gapping requests from the SCP.

Processing Functions
1. The SCP sends the SSP a request for ACG when it detects an overload condition at a particular destination.
2. The SSP initiates ACG as soon as it receives the ACG request. The mechanics of ACG Control are described in detail in Chapter 6, "GNS Application Description". Different levels of gap interval tables are used, depending on whether ACG is initiated manually or automatically.

12.3 Administrative Units in SSP

Trigger Table Administration
This unit administers the trigger table.

Processing Functions
1. Perform initial table load.
2. Provide the trigger indicators and the destination address to which the queries are sent.

    The following parameters are required:
    - trigger indication
    - CCS7 destination address
    - GTT indicator
    - primary or backup.
3. Perform the following functions:
    - CREATE   Make an entry in the table
    - MODIFY   Edit field within a table entry
    - CANCEL   Cancel a table entry
    - DISPLAY   Display a table entry.

    Note: Table entries are indexed using a key.

    These functions are usually entered from a local terminal connected to the SSP, or remotely from an Operations and Support System center.

Private Numbering Plan
This unit accepts dialog input concerning numbering plan definitions. It also generates or modifies the internal tables, databases, or data modules used by other PVN components.

Processing Functions
The network operator can specify:
- For each area wide business group, the digits to be used to indicate intra-switch calls or calls between locations.
- For calls to other switches, prefix digits to be outpulsed, and the routes to be used for each remote location.
- For PVN stations served directly by an SCP, that on calls to certain locations, a query be made to the SCP for routing information.
- That on calls to numbers assigned as network number calling numbers, a query be made for routing information.
- A numbering plan for use by switched data service users when calling each other.
- Certain lines or trunks as dedicated PVN lines or trunks.
- A prefix assigned to indicate a PVN call for a business group customer should be treated as a PVN call.
- A service code that may be used on certain lines (such as a **PBX** line as non dedicated PVN lines) to access a PVN. Translations indicate whether this code should be accepted from a line.
- Certain lines within a business group as PVN direct connect lines. Originations on such lines should be treated as PVN calls.
- For incoming business group tie-trunks, whether all calls on the tie trunk should be treated as PVN calls or if the calls received should be treated the same as calls originating from stations within the business group.

- Incoming PVN access tie-trunks on which the final digits are defined as a private network traveling class mark. Station group designators are identical to the traveling class marks and should be treated in the same way.
- Directory numbers to a PVN customer to be used for remote access.
- That calls to certain directory numbers should result in a query being made to an SCP database.
- In the SSP translations (for customers served by both SSP switches and non-SSP switches) PVN stations that are restricted from receiving non-PVN calls. The only calls to be completed to these lines are those received with a PVN indication.

12.4 Functional Units in SCP

A example of program layout is given in Figure 41. It includes the PVN application's functional and administrative units.

Dispatcher
As well as functions offered by the platform, the dispatcher performs some internal scheduling and dispatching. They may be centralized or distributed in the application program.
For the messages coming from the SSP or SMS:
1. Read the inquiry message from the application platform inquiry queue.
2. Check the status for congestion control.
3. Analyze and process the header/SCCP control data of incoming message.
4. Analyze and process the TCAP control information included in the message.
5. Save the header/SCCP control data and TCAP information.
6. According to the input data, dispatch the proper message handling unit.
The messages coming from the operator interface (MVS or node manager) are handled separately.

SSP-SCP Transactions
Several units handle the incoming SSP to SCP messages. The following is a description of the processing functions of an internal PVN call initialization dispatched after analysis of an incoming TCAP message (Invoke, Provide Instructions, or Start).

Processing Functions
1. Check and validate the input data of the message. If invalid, log or count the attempt,and prepare and send a refusal message.
2. Create a command for caller record search on the database (including the file name, key, input/output area, access parameters, and return code area).
3. Issue a database read (the actual read of the database is performed by the application platform).
4. After completion of database read, check and analyze the delivered return code.

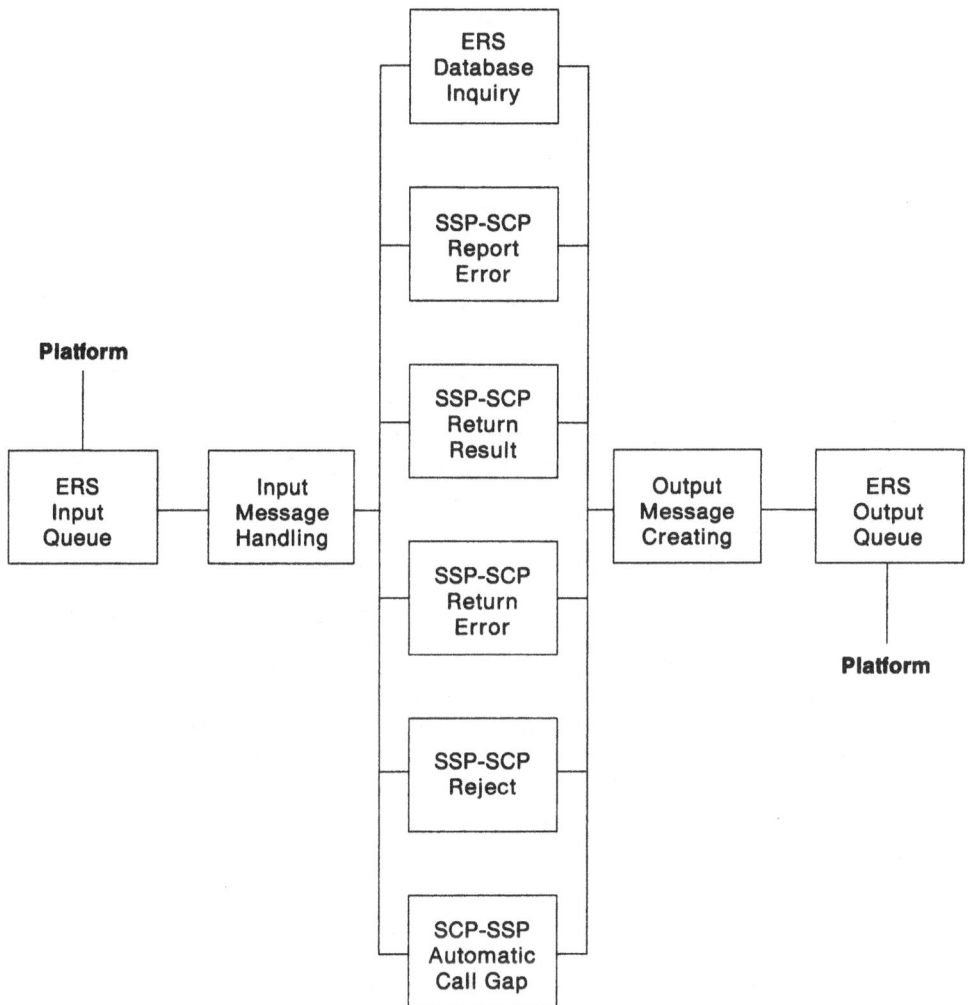

Figure 41. PVN Application Functional and Administrative Units

5. Create a command for a called party record search on the database (including the file name, key, input/output area, access parameters, and return code area).
6. Issue a database read (the actual read of the database is performed by the application platform).
7. After completion of database read, check and analyze the delivered return code.
8. Process the data of the records received to determine the real (target) number, considering time of day, day of week, screening, and other parameters. Determine the characteristics of the required connection. If a connection is not allowed, log or count the attempt, and prepare and send a refusal message.
9. Update counters used for resource control and statistics.
10. Create and format the data part of the output message.

11. Restore the header or control data, and TCAP information. Determine the final header or control data, and TCAP control information.
12. Create an output message consisting of header or control data, TCAP control information and the data part:
    TCAP (Invoke / Connection Control / Connect).
13. Write an output message on the application platform CCS7 output queue.
14. Update the counters used for traffic measurements and statistics.
15. Write log information if required (journaling).
16. Free resources (housekeeping) and return.

There are units that support the following messages and transaction types (they are not described in detail):

1. Internal PVN call. The call is initiated by a member of a PVN business group to another member, using the in-net dialing plan:
    TCAP (Invoke / Provide Instructions / Start)
2. Outgoing Call. The call is initiated by a member of a PVN business group to a PSN subscriber:
    TCAP (Invoke / Provide Instructions / Start)
3. Remote access call. The caller uses remote access dialing and must provide an access code:
    TCAP (Invoke / Provide Instructions / Start)
4. Connection acknowledge. The SSP has obtained a connection:
    TCAP (Return Result)
5. Connection failed. The SSP has not been able to obtain a connection:
    TCAP (Return Result)
    TCAP (Reject)
6. Call completion. The SSP has been asked to report the end of a PVN call to free the network resources:
    TCAP (Invoke / Send notification / Termination)
7. Authorization Code entry. The SSP sends the required additional digits:
    TCAP (Invoke / Provide Instructions / Additional)
8. Translation of a PVN number into a real number:
    TCAP (Invoke / Provide Instructions / Start)
9. Special Access:
    TCAP (Invoke / Provide Instructions / Start)
10. Operator support (display dialed number, translation, codes) when helping to establish a call:
    Any of the TCAP messages mentioned in points 1-9 above.
11. Network audio conferencing. A three-party call is established in the network:
    Any of the TCAP messages mentioned in points 1-9 above.
12. Switched data services. A call involving a data terminal is initiated:
    Any of the TCAP messages mentioned in points 1-9 above.

There is also a unit that monitors network resources and rebuilds an image of the PVN status by asking the SSPs for the active connections on each route. The normal mode is used to check that no messages have been lost between the SSP and SCP. It operates with a low priority and keeps the resource counters accurate. Recovery mode is used to recreate the resource counters after system interruption.

Interfaces to Application Platform and Queues
1. CICS Temporary Storage Read Next command to read input message from the application's inquiry queue.
2. CICS Temporary Storage Write command to write output message on the application's CCS7 output queue.
3. CICS EXEC LINK facility to access a database record and to write in application journal.

Data Structure
The data used by PVN application can be classed as:
- Information on network topology and group parameters for all PVN groups. To improve performance, parts of the file or the whole file are kept in main storage.
- Individual semi-permanent information which is kept on disk. One record is read via keyed access from the database which is a VSAM KSDS data file. The key for record search is built using either the dialed (virtual) subscriber number(which is supplied in the data part of the incoming inquiry message) or ANI-calling number, or a PIN.
- Transient information (such as resource counters). To improve performance, parts of the file or the whole file are kept in main storage.
- Records on the journal file which is a VSAM ESDS data file (write only).

Input and Output Data
1. The TCAP part of the incoming message contains coded information on the actions required by the application program (such as Provide Target Number).
2. The data part of the input message is either related to the originating party (such as the user group, number, or ANI) or consists of the dialed (virtual) subscriber number.
3. The data part of the output message may be connection information (for example, the translated number, or the operator number) or a number of digits to be collected.
4. The TCAP part of the output message supplies coded information for the functions required by the SSP.

SMS-SCP Transactions
Several units handle the incoming SMS to SCP messages. The following is a description of the processing functions of a unit handling a PVN database update for the class of service of an individual member of a PVN group.

Processing Functions
1. Check and validate the input data of the message.
2. Create a command for the selected record search on the database (including the file name, key, input/output area, access parameters, and return code area).
3. Issue a database update (the actual update of all PVN database copies on the different SCP processors is performed and controlled by the application platform).
4. After completion of the database update, check and analyze the delivered return code.
5. Create and format the data part of the response message to the SMS.

6. Restore the SMS-SCP header and determine the final header for response.
7. Create a response message consisting of an SMS-SCP header and data part.
8. Write a response message on the application platform SMS output queue.
9. Update counters used for traffic measurements and statistics.

There are units that support the following messages and transaction types (they are not described in detail):

- Update of system-wide definitions
- Update of PVN-group definitions
- Update of PVN group configurations
- Update of individual PVN user definitions
- Individual Delete
- Individual Create
- Global Delete
- Global Create
- Report or Display.

Some changes may be incompatible with an on-line application (for example mass updates or system parameters). The update should be performed off-line and followed by a global change of the database at application restart.

Interfaces to Application Platform and Queues

1. CICS Temporary Storage Read Next command to read the input message from the application's update queue.
2. CICS Temporary Storage Write command to write the output message on the application's SMS output queue.
3. CICS EXEC LINK facility to update a PVN database record and write the PVN update application journal.

Database Access

1. Update a subscriber record via keyed access on the PVN database which is a VSAM KSDS data file. To improve performance, parts of the file or the whole file may be kept in main storage. Control of the storage medium is done by the application platform.
2. The platform also ensures the proper update (in case there are several copies of the PVN database replicated on different SCP processors).
3. For improved performance an update transaction may consist of a batch of updates affecting several database records.
4. Write a record on the PVN journal file which is a VSAM ESDS data file.

Input and Output Data

1. The SMS-SCP header of the incoming update transaction contains information on the actions required by the update application program (for example, a batch of updates).
2. The data part of the update message consists of individual subscriber data to be added, changed or deleted (such as routing parameters like time of day, day of week, class of service, alternate numbers, screening data to handle calls, depending on the call originating area) or PVN network data (such as the number of dedicated trunk lines on a route).

3. The data part of the response message includes information on successful (or unsuccessful) record updating.
4. The SMS-SCP header of the response message contains additional control information for the SMS.

12.5 Administrative Units in SCP

Application Initialization
When the application program is started, this unit checks that all the required resources are available, and initializes the dialog with relevant partners.
1. Get main storage for resident tables and queues.
2. Open the PVN databases.
3. Read the data that must be kept.
4. Initialize the queues.
5. Tell the relevant partners that the application is now running.
6. Tell the operator if something is missing to run the application.

Miscellaneous Units
There are routines to perform specific functions for functional and administrative units, for example:
*   **Congestion Control**
    A periodic congestion control routine can detect overload (for example an excessive response time of the SCP) and take action to help alleviate the overload. The action taken depends on the level of overload. Automatic Call Gapping (ACG) is one of these actions, but can also be triggered by an operator command. Administrative activity is affected first, and if the overload condition continues, processing activity is affected.
    The congestion control routine has the following functions:
    1. Determine the values of traffic counters for the last period.
    2. Analyze the database load for this time period.
    3. Analyze the application input queues for unprocessed PVN requests.
    4. Analyze the response time counters.
    5. Check the number of application clone tasks for the PVN application.
    6. Process all these parameters to determine the load:
        If normal load situation; no action (return). If overload situation or operator initiated ACG, the processing of input parameters provides the overload level and determines the contents of the message to be sent.
    7. Create and format the data element of the TCAP message:
        − TCAP (Invoke / Network Management / ACG)
    8. Update counters for ACG control and statistics.
    9. Call TCAP Output Message Creating function.
*   **Configuration Services**
    Some actions taken in the functional units are determined by the topology of the user groups PVN. Specific routines handle that for all general purpose units.

- **Traffic Reporting**

   The functional units maintain measurements (peg counts) and log the exceptions (if required) for all the general purpose units.

   Logged exceptions are routed to the SMS either immediately or on a periodic basis (on SMS request). At a predetermined time, the measurements are sent to the SMS and the counters are reset to zero. The period is 30 or 60 minutes, according to the type of counter. This unit also includes the error recovery and initialization counter procedure.

- **Database access utilities**

   Some actions taken in the functional units depend on correlated data in the database which may require non-trivial operations (such as table look-up or alternate choices). These specific routines handle these non-trivial processes for all the general purpose units.

Part 6. Area Wide Centrex

This part is a description of the Area Wide Centrex service. It does not contain an application description.

This part is AWC-specific; it adds to, and does not replace "Part 1. Overview of the Intelligent Network" of this report.

Chapter 13. AWC Service Description

13.1 Overview

Basic Service

Centrex service provides business features via a central office. Callers from locations served by that office and belonging to a business user group can use a centralized numbering plan, abbreviated dialing and PBX features like call forwarding and conference calls, as if they were connected to individual PBX systems. Centrex acts like an individual system for each group.

Area Wide Centrex (AWC) expands the Centrex service across a number of exchanges.

Background

Some countries' networks offer a Centrex service with special telephone functions. The Centrex system is a physical part of a central office or exchange and provides functions similar to PABX. The user can define a special numbering plan to the service subscriber.

As all stations of a business user group are directly connected to the central office, a Centrex system may be restricted to a local area. Using Area Wide Centrex (AWC) the area can be expanded; stations connected to a Centrex system receive calls from, and access local users via the public switched network through AWC functions. This can be helpful and economical for decentralized national organizations.

Benefits

AWC provides inter- and intra-exchange Centrex features with the following characteristics:

*   Centralized numbering plan

    Any station within a business group can call another using a fixed length number (shorter than the network address, because only one business group must be covered by this number).

*   Feature transparency

    A subset of Centrex features are active across the whole area. The station at a given exchange notices no difference when handling calls even when they are received from, or directed to, stations on other exchanges.

- Centralized attendant service
  The attendant for the whole area with numerous exchanges may be located at one exchange.
- Central management
  Changes applicable to a whole area with several exchanges can be performed from a central point.

The benefits of AWC to the service subscriber are:

- Minimal capital expenditure to obtain sophisticated business features comparable to those available from PBX services.
- No need for dedicated facilities
- Maintenance, administration and network planning is provided by the Telco.

The benefits of AWC to the network operator are:

- Better exploitation of network hardware capabilities
- Improved market position in a competitive environments, by offering an alternative to PBX.

13.2 Functional Description

General

AWC stations are connected to a SSP which is an end exchange or tandem exchange. When a service user A makes an AWC call, the SSP of the originating station checks the caller's Class of Service (COS), and triggers a query to the AWC database located in the SCP if the call cannot be handled locally.

There are special numbers which always trigger an SCP request; this query is transmitted to the SCP via the CCS7 network and STP nodes. The AWC application program in the SCP reads a record from the AWC database using the:

- Dialed number
- Originating number to identify the user group (service subscriber)
- Originating Class of Service (COS).

This information is included in the query. The AWC application program translates the dialed number to a real number according to the defined numbering plan for that service subscriber. This target number and service logic instructions are sent to the SSP via the CCS7 network and STP nodes. The SSP and SCP communicate using the TCAP protocol.

The SSP completes the call setup and routes the call through the public switched network to the appropriate destination of the same user group at another exchange.

The AWC service can be customized so that the service subscriber uses the SMS via the network operator, or directly by a data terminal allowing fast and flexible service. Transaction processing between the SMS and the SCP transfers the subscriber specified data to the SCP(s) where it is used for AWC calls.

Table 5 shows current Centrex features and enhancements for Area Wide Centrex features.

Not all Centrex features are suitable for AWC. Where it is unclear whether they can be extended across several exchanges with moderate effort, a question mark is inserted.

Table 5 (Page 1 of 2). Centrex and Possible Area Wide Centrex Features		
FEATURE NAME (According to LSSGR)	**Expandable to AWC**	
	Today	**Future**
Business Group Dialling Plan	X	
Distinctive Ringing/Call Waiting Indications	X	
Special Intercept Announcements	?	
Virtual Facility Groups	?	
Centrex Complex	?	
LINE FEATURES:		
Intercom Dialling	X	
CALL TRANSFER FEATURES:		
Call Transfer - Internal only (Group only)	X	
Call Transfer - Individual All Calls	X	
CALL HOLD FEATURES:		
Call Hold		
Add On Consultation Hold-Incoming Only	?	
Call Park	?	?
CALL FORWARDING	X	
CALL PICK UP FEATURES:		
Call Pick Up	na	
Directed Call Pick Up	na	
Secretarial Intercept		X
Speed Calling	X	
Three Way Calling	X	
Trunk Answer Any Station	?	?
Do Not Disturb		X
MULTILINE HUNT GROUP (MLHG) FEATURES:		
Multiline Hunt Service	X	
Preferential Multiline Hunting	?	
Uniform Call Distribution for MLHG's	X	
Queuing for MLHG's	X	
Delay Announcement for MLHG Queues		
Series Completion	X	
PRIVATE FACILITIES ACCESS AND SERVICES:		
Tie Facility Access	?	
Dial Access to Private Facilities	?	
Facility Restriction Level (FRL)	?	

FEATURE NAME (According to LSSGR)	Expandable to AWC	
	Today	Future
Table 5 (Page 2 of 2). Centrex and Possible Area Wide Centrex Features		
Travelling Class Marks (TCM)	?	
Radio Paging		X
Loudspeaker Paging		X
Code Calling		X
Recorded Telephone Dictation		
ATTENDANT FEATURES (Centralized Attendant):		
Call Origination	X	
Call Termination	X	
Call Extension	X	
Call Swap		
Call Split		
Extended Disconnect		
Attendant Conference		
Attendant Camp-On		X
Indication of Camp-On		X
Attendant Line Busy Verification		X
Attendant Tie Trunk Busy Verification		X
Attendant Call Through Test on Tie Trunk		X
Attendant Emergency Override		X
Attendant Access to Code Dialling		X
Dial Through Attendant	X	
Station Billing on Attendant Handled Calls		
Night Service	X	
ATTENDANT CALL DISTRIBUTION FEATURES:		
Multiple Position Hunt With Queuing		
Hunting for Multiple Attendant Positions		
Attendant Console Queuing		
Call Forwarding Between Attendant Consoles		
INDICATORS FOR ATTENDANT CONSOLE:		
Trunk Group Busy Indicator		X

Service User's Perspective

Area Wide Centrex provides transparent service across distributed user locations, that is, the user perceives no difference in feature operation or level of service compared with service from one local exchange.

Incoming Calls

- Central Attendant Service (CAS). Attendants at a central location can handle calls for several different user groups and at different exchanges.
- Multiline Hunting. Hunting groups within one exchange may overflow to another hunting group at another exchange to avoid overload or congestion of agents or attendants.
- Internal Call Routing. Features are selected by the customer and introduced into the network by software:
  - Call forwarding diverts a call to another line, thus facilitating automatic forwarding when the station called is busy or does not answer within a given time. The forwarding may also be activated for the free-status of a station leading to an unconditional call forwarding to another station in the area.
  - Call transfer directs a call to another line. Three-way communication is possible. If a receiver on the other line does not answer, the call is transferred back, or transferred to the group attendant.
- Night Answer answers incoming calls when the group attendant is unavailable. Calls may be diverted to any station in the area.
- With Universal Night Answer, special tone or chime announces an incoming call. Any station of the group within one exchange can answer the call by entering an access code.
- Do Not Disturb prevents incoming calls from ringing. It may be over-ridden from stations with the appropriate class of service. This function is also known as incoming barring.

Outgoing Calls

- Speed dialing. Frequently called numbers can be entered into the system and are then called using an abbreviated number. Abbreviated dialing instructs the system to dial the full number. There may be "common" speed dialing lists which are common to the whole business user group and which can be updated only from the SMS. There are also individual lists pertaining to an individual station.
  If the station has a private list the service user may create a private directory of his own frequently called numbers. The station may change the numbers at any time.
- Conference Calls: A call-in-progress can add parties to the conversation by entering a code and the number of the party to be added. More than one party may be added.
- Usage limitation (Call Barring):
  - Class of service restrictions prevent access to specific features from specific stations
  - Prefix restrictions limit access to specific prefixes such as "dial-a joke", or toll calls from specific stations.

Any call from a Centrex or AWC station is checked for these restrictions before further call processing. Calling procedures and additional functions within a local user group of a Centrex remain unchanged.

The main benefit of AWC to the service user is the convenience of having PBX-type features over several locations, and the ability to retain a number when moving to new locations.

Service Subscriber's Perspective
If the AWC service subscriber has members of his organization in remote locations, he can adapt his communication services to his organizational requirements.
Remote lines can be integrated into the company numbering scheme and features assigned according to function.
Modifications can be made quickly and easily through the SMS by:
* Asking the network operator by letter or phone to make the necessary changes
* The service subscriber accessing the SMS and changing the SMS database entries within the permitted boundaries. This is controlled by menu and security restrictions.

Network Operator's Perspective
The benefits of AWC to the network operator are:
* Service revenue
* Balanced traffic on local access lines
* Controlled management and easier administration of services.

Relation to ISDN Supplementary Services
AWC is independent of ISDN. In a mixed environment ISDN and non-ISDN exchanges may belong to the same AWC business group. Service users belonging to an ISDN exchange, however, will profit from the ISDN generic functions which allow a higher degree of feature transparency.

13.3 Standards

There are no dedicated AWC standards yet. However the following are AWC-related and are the basis of this service description:
* *TCAP Message Format for 800 Service*, Bellcore LSSGR TR-TSY-000064
* *Feature Transparency for Multi-Location Business Groups*, Bellcore LSSGR TA-TSY-000404

13.4 Service Interaction

Service User

Class Of Service: All interactions are controlled by the station's Class of Service (COS). The COS is tailored so that an individual station receives the level of service required by the service user at his station.

Making calls: For internal calls (intercom dialing), the user dials the desired station number according to the business group numbering plan. This number is shorter than the network address number. For outside numbers (in AWC and Centrex this may be a station at the same exchange, but belonging to another business group, or no group at all) the entire network address number must be dialed, eventually preceded by a DOD prefix.

Receiving Calls: The sources of an incoming call can be signalled by distinctive ringing. The calls may be directed to a single station or a group of stations. Multiline hunting, uniform call distribution, and queuing may be used to select an individual station.

Activating Features: The user may activate different features by dialing a prefix or suffix during a call (for example call forwarding or initiating a three-way call) or by dialing a prefix from idle status (for example, call pick-up and night service answer).

ISDN Stations: AWC stations using ISDN exchanges profit from the whole ISDN function set. When all exchanges used to provide AWC to a customer are ISDN, service enhancements are possible. ISDN will have an impact on AWC, for example, for future ISDN-only topologies, a new AWC feature set may be defined.

The service subscriber may provide a user's manual. Depending on the instrument the user may receive guidance on his display or lighted buttons indicating which feature may be activated at the current state of call. For an ordinary telephone, different tones or announcements guide the user. Depending on the station's class of service, the user may enter speed dialing numbers and forwarding numbers.

Service Subscriber
When interacting with SMS, the service subscriber may enter the following data:
- Service subscriber (customer) number
- Station user's name and address
- Station's physical location
- Business group
- Security passwords
- Feature activation prefixes
- Routing parameters (for example time of day, day of week)
- Line number
- Class of service
- Real number
- Instrument type and characteristics
- Recorded announcements.

The attendant console enables the subscriber to directly change data and get special billing and traffic data. The attendant console also provides information and call connection services for AWC subscribers and for callers outside the subscribers user group.

Network Operator
Refer to Chapter 2, "Functional Characteristics Common to Selected IN Services".

Access Instrument

Service user: The AWC can be used from any station (POTS or ISDN) belonging
to the group. ISDN stations are more convenient (for example, display guidance
and dedicated illuminated buttons). For compatibility, the AWC feature set is
common to any service user.

13.5 Billing

The billing requirements are highly country specific, and may include:
* Lump sum billing for any Centrex/AWC call within a business group
* Lump sum billing but for given geographical areas
* Individual billing for AWC calls only.

SSP
The originating exchange (SSP) must be able to generate AMA records for all
TCAP AWC calls using the billing information received from the SCP database.
The billing indicator field is a mandatory field in the response, and therefore
requires that an AMA record be generated.

An AMA record should be made for all calls answered. The terminating
exchange should be able to bill the AWC calls on a service subscriber basis. Billing
information may also be provided at the attendant console.

SCP
No requirements exist for AWC billing functions in the SCP.

SMS
Refer to Chapter 2, "Functional Characteristics Common to Selected IN Services".

13.6 Service Logic

13.6.1 Distribution

SSP

The SSP database should contain originating triggers to determine whether or not special routing is required.

For example, if an AWC station goes off hook the SSP knows the user group to which the station belongs, and the station's class of service (COS). The SSP also knows what numbering plan is to be used and what features are allowed. Using the AWC-number dialed, the SSP can detect whether or not the number belongs to a station served by the SSP. If not, a message to the SCP is triggered containing the dialed number, the number and the COS of the calling station, and the request for translating the called (target) AWC number to a network address.

The SSP completes the call setup using the translated network address. IN service logic must be able to coexist with Centrex logic.

SCP

The SCP contains service logic enabling it to translate an incoming station number to a real destination number, depending on the origin and the service subscriber's numbering plan for that business group. The SCP also evaluates the class of service for both stations to check whether the SSP's request can be handled.

For example, when a station requests call forwarding to a station at an SSP with no appropriate COS to receive outside calls, the SCP should deny the forwarding request, depending on the call's origin.

The SCP must contain service logic to handle exceptions such as overload situations, database error situations and cases where there is no number assigned.

Triggers

The trigger table in the SSP is administered by the network operator and describes which part of the call data will trigger a query to the SCP and at which point in the call the query is started. The triggering data for AWC is:

- Origin number of a user, indicating he is a member of a specific user group, and calling an AWC station not belonging to the originating SSP.
- Dialed prefix or suffix to access special features.
- Class of Service (COS) of originating station.

The SCP query is started as soon as the information (the called number) is received, and the SSPs verification of the COS is positive.

13.6.2 Functional Flow

SSP-SCP

The following description of the inter-exchange activities controlled by the SCP assumes that no transient data related to the different call phases is stored in the SCP (with one exception: uniform call distribution across several SSPs). Transient data defines temporary conditions relating to a call or a station, and conditions the routing of the call. It is managed in the SSP of the station to which the data belongs. This reduces the SCP's load, because only translation and dispatch tasks are performed by the SCP.

The functions described below can be extended from Centrex to Area Wide Centrex. The descriptions show the operation of AWC, and the messages for call setup. A common figure with three stations A, B, and C at three SSPs but all belonging to the same business group is used. Outside calls from an outside line or another business group are shown as a flash. The TCAP messages used are:

Component Type	Family Name	Specifier	Direction: SSP SCP
Invoke	Provide Instructions	Start	------->
Invoke	Connection Control	Connect	<-------*
Invoke	Network Management	Automatic Call Gapping	<-------
Invoke	Send Notification	Termination	<-------
Return Result			------->
Invoke	Caller Interaction	Play Announcement	<-------
Return Error			<-------
Invoke	Procedural	Report Error	------->
Reject			<------>
Invoke	Network Management	Update	<-------

(* In the U.S. only. CCITT requires Return Result.)

For the inter-office signaling between SSPs, standard ISUP protocol messages are used. The minimum ISUP messages used for call setup and release required for AWC are:

- Initial Address Message (IAM)
  IAM is sent to the exchange at the terminating end of a circuit to initiate circuit seizure for an outgoing call. The IAM includes address information and other information about the routing and handling of the call.
- Address Complete Message (ACM)
  ACM is sent in the opposite direction to the IAM, and indicates that all address signals required for routing the call to the called party have been received.
- Answer Message (ANM)
  ANM is sent in the opposite direction to the IAM, and indicates that the call has been answered.
- Facility Request Message (FAR)
  FAR is sent in either direction to request activation of a facility.
- Facility Accepted Message (FAA)

- Facility Accepted Message (FAA)
  FAA responds to a FAR indicating that the requested facility has been activated.
- Facility Rejected Message (FRJ)
  FRJ responds to a FAR indicating that the requested facility has been rejected.
- Release Message (REL)
  REL is sent in either direction and indicates that the circuit identified in the message is being released. The exchange receiving this message should release this and connected circuits.
- Release Complete (RLC)
  RLC is sent in either direction in response to the REL message when the identified circuit is released.

Business Group Dialing Plan

The business group dialing plan is tailored to the individual AWC service subscriber's requirements. AWC subscriber stations call each other across multiple exchanges without using a complete network address. The length of the number depends on the size of the business group. Station numbers may be repeated in other business groups. All numbers are stored in the SCP and have a real network address across all AWC exchanges assigned. The local SSP has assigned network addresses only for station numbers attached to that SSP.

Examples: A functional flow is described for the following AWC features:
- Intercom Dialing
- Distinctive Ringing
- Call forwarding
- Speed Calling
- Multiline Hunting
- Attendant service
- Night service.

Intercom Dialing

1. SSP-A transmits a translation request to the SCP containing A's identification (A number and COS) and the B number.
2. The SCP checks whether A's COS is allowed to connect with B's COS. If connection is allowed, the SCP returns the A-number and the B-number together with B's network address. Connections may be denied because the stations are incompatible or because the B-station lacks authorization.
3. SSP-A then requests the connection to SSP-B, using the initial address message of the ISUP protocol.
4. When a connection to Station B is established (so that station B is ringing), SSP-B returns the ACM message to SSP-A.

Distinctive Ringing

Distinctive ringing gives the service user information about the origin of a call, so he can respond appropriately. Each SSP can use the "Nature of connection indicator" of the IAM message to trigger distinctive ringing. Depending on the

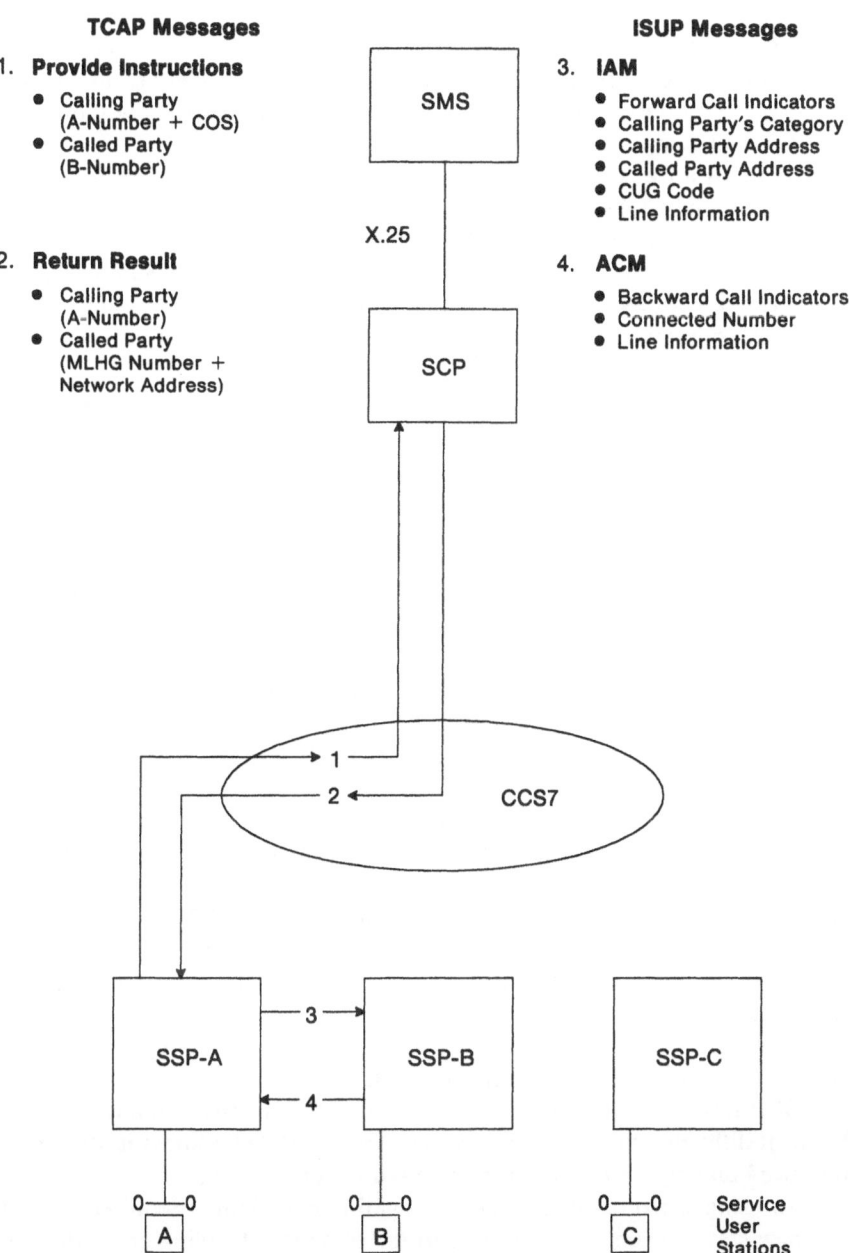

TCAP Messages

1. **Provide Instructions**
   - Calling Party
     (A-Number + COS)
   - Called Party
     (B-Number)

2. **Return Result**
   - Calling Party
     (A-Number)
   - Called Party
     (MLHG Number +
     Network Address)

ISUP Messages

3. **IAM**
   - Forward Call Indicators
   - Calling Party's Category
   - Calling Party Address
   - Called Party Address
   - CUG Code
   - Line Information

4. **ACM**
   - Backward Call Indicators
   - Connected Number
   - Line Information

Figure 42. AWC Intercom Dialing Functional Flow

available cadences of the SSP and local regulation the AWC subscriber may define how to signal which call.

Call Forwarding

If station A calls station B, station B is call forwarded to station C. The conditions for call forwarding may vary; for example, when there is no answer within a specified interval, when Station B is busy, or immediately without ringing (to cover temporary user absence).

1. SSP-A sends a translation request to the SCP containing A's identification (A-number and COS) or network address, and the B-number.
2. The SCP checks whether A's COS is allowed to connect with B's COS. If connection is allowed, the SCP returns the A-number, the B-number and B's network address. The same is done with the network address of the outside station when receiving a call from outside (individual non AWC call) instead of an A-station call.
3. SSP-A then requests the connection to SSP-B using the initial address message of the ISUP protocol.
4. SSP-B determines that Station B is call forwarded. SSP-B returns the REL message with redirection information including a redirection indicator parameter set to "Call Forwarded" and the redirection number (Station C's address).
5. SSP-A acknowledges the release message with the release complete message.
6. SSP-A then requests connection to SSP-C via the IAM.

Speed Calling

This function lets an AWC service user store a list of short codes that can be dialled instead of longer numbers. Depending on the required dialing speed, there may be one target assigned per button at the keypad. An "exit" button is assigned to dial normal numbers.

The speed dialing lists are located in the SSP serving the stations assigned to that list. Outside or non-Centrex numbers are stored in the SSP according to its Class of Service (COS); there is no SCP involvement. When a target number is to be assigned to a position in the speed dialing list, the SSP first checks the COS. If the COS is compatible, the target is known by the SSP, and the list is defined as updatable from the station itself, the SSP directly enters the network address to that list position.

1. Assuming the address added to the list is at SSP-B, the real network address must first be obtained from the SCP.
2. The SCP returns the network address, as an AWC intercom dialing call between different exchanges, except that the translated address is stored in the speed dialing list instead of being used to setup a call.
3. To call using speed dialing, SCP interrogation is no longer required.

When a new station number is assigned to the same speed calling code, the list entry is replaced, and the previous network address is erased. A new network address is obtained from the SCP if the new number does not belong to that SSP.

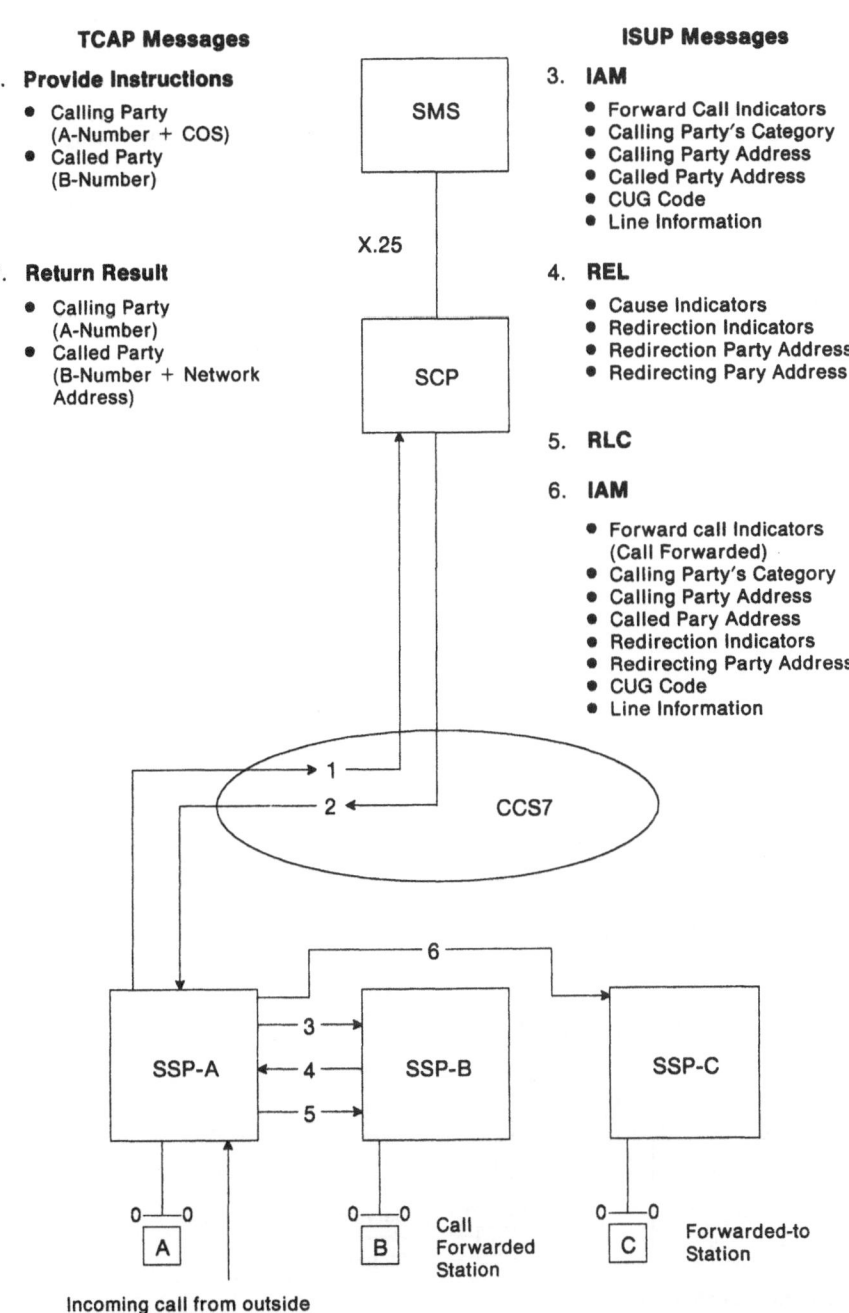

TCAP Messages

1. **Provide Instructions**
   - Calling Party
     (A-Number + COS)
   - Called Party
     (B-Number)

2. **Return Result**
   - Calling Party
     (A-Number)
   - Called Party
     (B-Number + Network
     Address)

ISUP Messages

3. **IAM**
   - Forward Call Indicators
   - Calling Party's Category
   - Calling Party Address
   - Called Party Address
   - CUG Code
   - Line Information

4. **REL**
   - Cause Indicators
   - Redirection Indicators
   - Redirection Party Address
   - Redirecting Pary Address

5. **RLC**

6. **IAM**
   - Forward call Indicators
     (Call Forwarded)
   - Calling Party's Category
   - Calling Party Address
   - Called Pary Address
   - Redirection Indicators
   - Redirecting Party Address
   - CUG Code
   - Line Information

Figure 43. AWC Call Forwarding Functional Flow

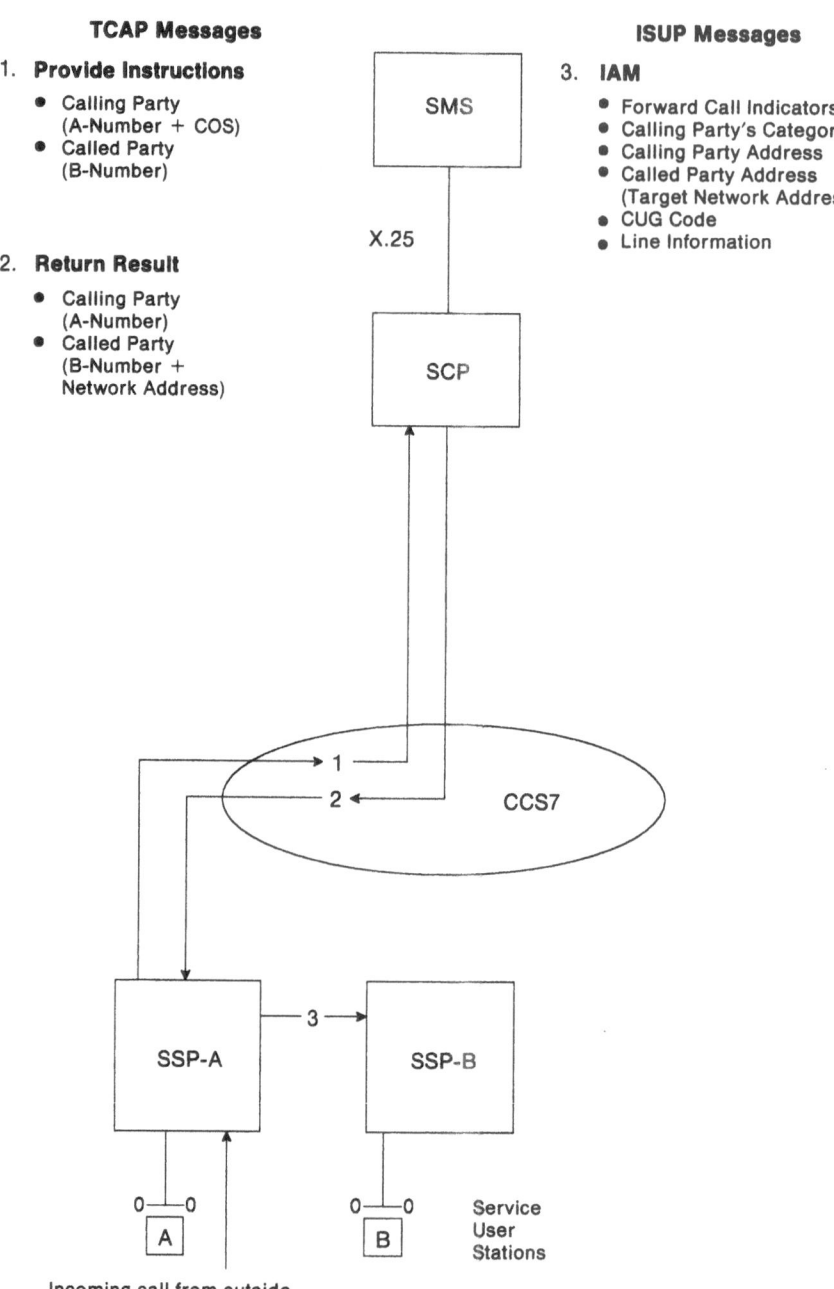

TCAP Messages

1. **Provide Instructions**
   - Calling Party
     (A-Number + COS)
   - Called Party
     (B-Number)

2. **Return Result**
   - Calling Party
     (A-Number)
   - Called Party
     (B-Number +
     Network Address)

ISUP Messages

3. **IAM**
   - Forward Call Indicators
   - Calling Party's Category
   - Calling Party Address
   - Called Party Address
     (Target Network Address)
   - CUG Code
   - Line Information

SMS

X.25

SCP

1

2

CCS7

3

SSP-A

SSP-B

A

B

Service
User
Stations

Incoming call from outside
or other business group

Figure 44. Speed Calling Functional Flow

Multiline Hunt Group (MLHG) Features

• **Multiline hunt group service (Distribution in SSP)**

Where a call originates from a business group station at one SSP, calling a MLHG at another SSP, is the only case where the SCP is really required. The typical situation for most calls is an outside call arriving directly at the SSP where the MLHG is installed. In this situation steps 1-3 described below do not take place, and the service runs without the SCP.

1. SSP-A receives an incoming call from outside (that is, a station belonging to another business group or a real outside line calling in) or a station calling a multiline hunt group for service by a single group number. If the desired group is not in SSP-A, the SSP-A asks the SCP to provide the MLHG address.

2. The SCP provides the MLHG network address.

3. SSP-A forwards the MLHG address to the SSP-B, with a request for connection.

4. SSP-B (where the desired MLHG is installed) selects an available station out of the group. There are different selection criteria:

   – Hunting in series completion: There is a fixed starting station address from which the next station, in fixed sequence, is selected. This places greater strain on the stations at the beginning of the search list.

   – Hunting with uniform distribution: The calls are presented to all available stations uniformly by searching the available stations cyclically. The workload is more balanced than with hunting in series completion.

   – Hunting with concentration: A queue of calls is presented to all stations simultaneously. The call is received by the station that accepts it first. This method gives the fastest caller response, but operators who work quickly will have a heavier workload.

   There are different stages for call queuing on multiline hunt groups. They all are extensively described in independent documents for MLHGs or ACDs (Automatic Call Distribution). What is important for this example is the overflow condition only, or the occasions on which the SSP rejects new calls for its MLHG.

   An overflow condition is defined by the service subscriber for each SSP MLHG. If an overflow condition is reached, the MLHG in SSP-B runs on an alternative address, pointing to the next MLHG of that AWC subscriber. SSP-B sends the network address of the next MLHG member with a REL message to the originating SSP-A. (In this example a MLHG in SSP-C)

5. SSP-A acknowledges the REL with a release complete message.

6. SSP-A makes another attempt to connect the call, this time using the MLHG address for the next member at SSP-C.

During peak traffic periods, steps 4 to 6 must be repeated; SSP-C would answer like SSP-B in step 4, and provide an alternative address for the **next** MLHG after overflow. By this method, multiline hunting may be built up across several exchanges.

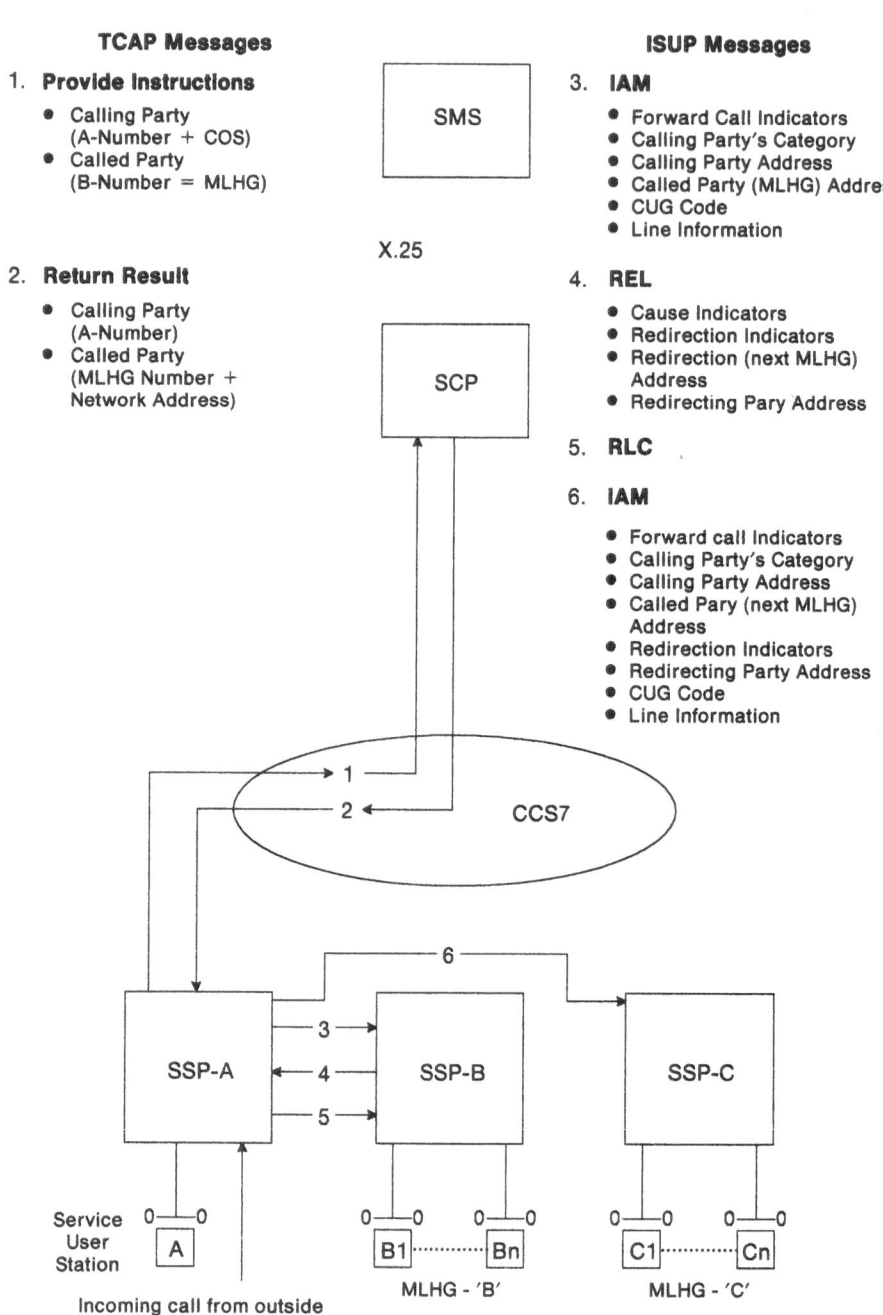

Figure 45. MLHG Distribution in the SSP Functional Flow

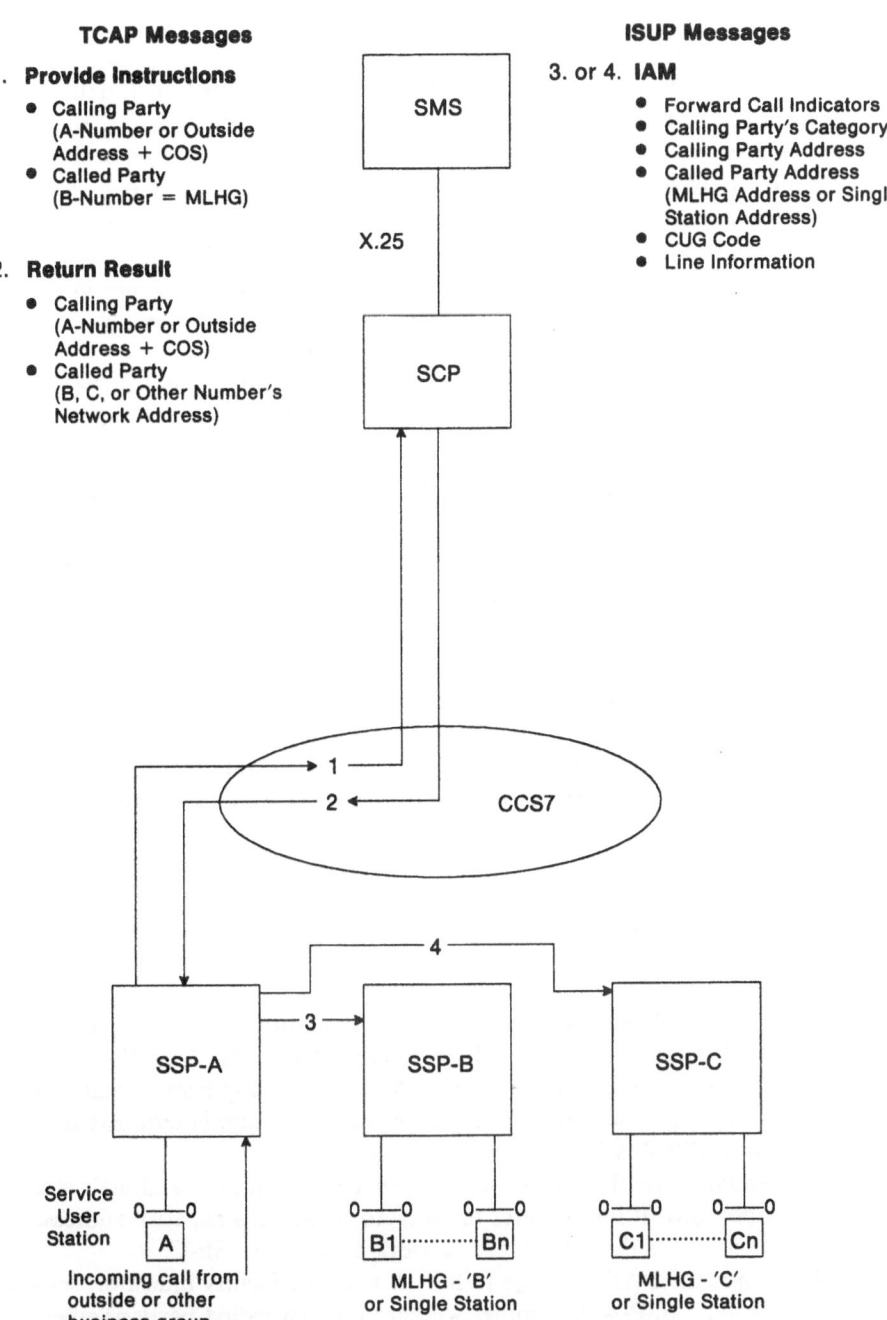

TCAP Messages

1. **Provide Instructions**
   - Calling Party
     (A-Number or Outside
     Address + COS)
   - Called Party
     (B-Number = MLHG)

2. **Return Result**
   - Calling Party
     (A-Number or Outside
     Address + COS)
   - Called Party
     (B, C, or Other Number's
     Network Address)

ISUP Messages

3. or 4. **IAM**
   - Forward Call Indicators
   - Calling Party's Category
   - Calling Party Address
   - Called Party Address
     (MLHG Address or Single
     Station Address)
   - CUG Code
   - Line Information

SMS

X.25

SCP

CCS7

1

2

4

3

SSP-A

SSP-B

SSP-C

Service
User
Station

A

Incoming call from
outside or other
business group

B1 ········· Bn

MLHG - 'B'
or Single Station

C1 ········· Cn

MLHG - 'C'
or Single Station

Figure 46. MLHG Distribution in the SCP Functional Flow

- **Multiline hunt group service (Distribution in SCP)**
  This MLHG service is located in the SCP. It allows a uniform call distribution across multiple exchanges, and is the only service in which transient data is managed in the SCP, and where all calls to the MLHG are passed via the SCP.
  1. SSP-A receives an incoming call from outside or from a station calling that multiline hunt group for service by a special single group number. Special calls, for example, from a special direction or source, may be routed to that MLHG. Even when the group exists in SSP-A, the SCP is interrogated to route the call to the next station according to the area wide uniform call distribution. This station may belong to the exchange if the originating SSP has stations belonging to that SCP-controlled MLHG.
  2. The SCP provides the network address of the next station in the sequence of uniform distribution. If the selected station belongs to SSP-A, the procedure stops here.
  3. and 4. If another station address is provided, SSP-A requests that the call be connected to that station at another SSP.
  Single stations in the SSPs may be MLHGs as previously described. The area-wide uniform distribution can then be considered a "super-distribution".

Attendant Features

The AWC **Centralized Attendant** feature lets users locate their attendants at one convenient location. Calls to the attendant originating within the service user group are routed to the SSP terminating the attendants via intercom calling

AWC Directory Number calls that originate outside the business group are routed to the SSP serving the attendants via a function similar to intercom calling. In both cases, the originating SSP in the service user's network knows that the call must be routed to an attendant. When the attendant is not collocated with the business group, the SCP is queried to obtain the attendants network address.

Like any AWC business group station, the attendant can originate and terminate AWC calls using the intercom dialing procedure.

Call Extension

1. The attendant of SSP-B can transfer a call to station C at SSP-C using a call transfer mechanism. The call could come from SSP-A or directly from SSP-B. If the call originates from SSP-A, it has already been handled by the intercom dialing procedure. In any case the connection is terminated at the attendant at SSP-B when this procedure starts.
2. The attendant puts the connection to be extended on hold and dials station C's number. SSP-B interrogates the SCP to obtain C's network address.
3. The SCP provides the station C's network address at SSP-C.
4. SSP-B sends the FAR message to SSP-A with the facility indicator, set to "call transfer". SSP-A determines whether the new called party number is locally terminated.
5. If station C is not terminated by SSP-A, SSP-A sends the FRJ message to SSP-B.
6. SSP-B then initiates a connection request to SSP-C.

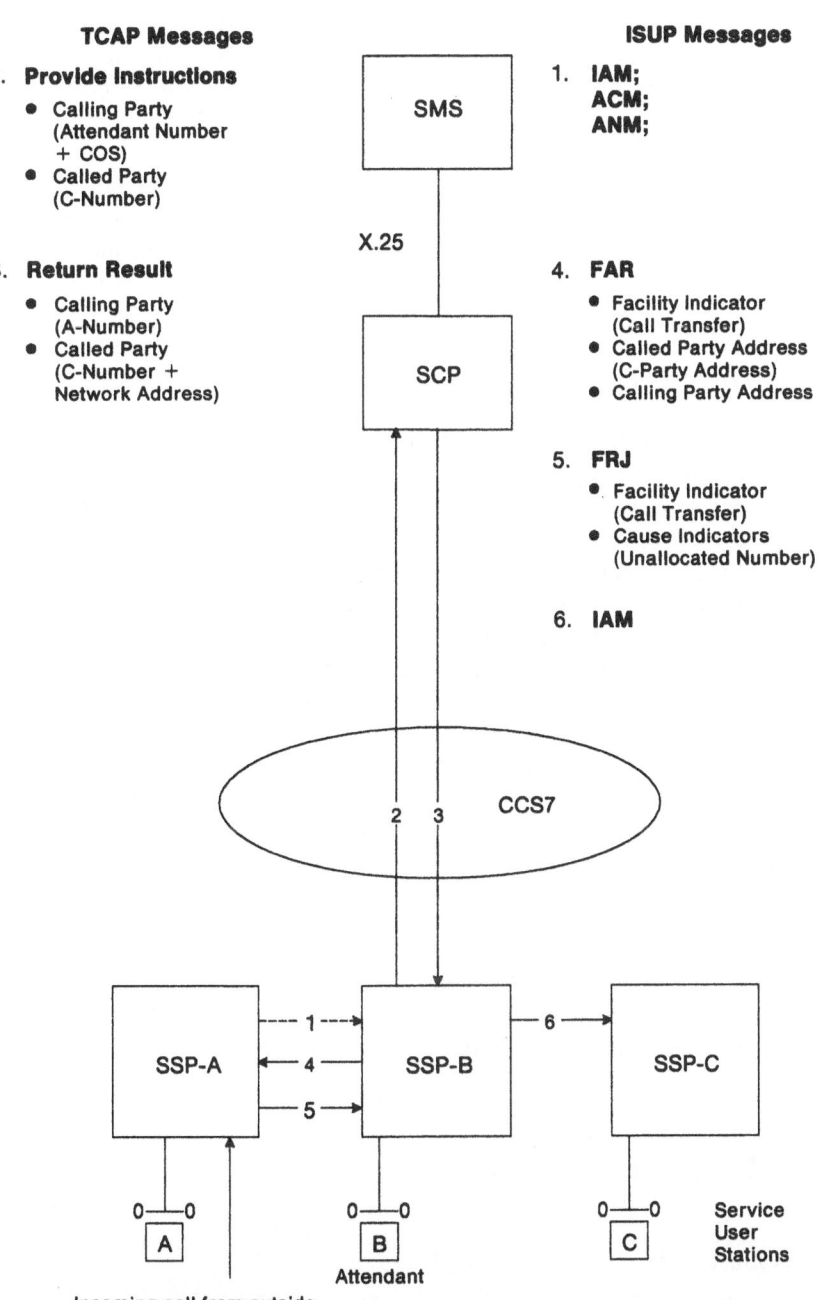

Figure 47. Call Extension Functional Flow

This procedure is not limited to attendants. Stations with the call transfer feature can also extend calls. Consideration should be made of the number of subsequent call transfers allowed on one call, as several links will be occupied for a single call.

Night Service

It is assumed for this description that the AWC subscriber uses attendants located at one exchange and night service stations at another exchange. There may be several night service stations in a multiline hunt group; this means that the multiline hunt group procedures apply at the end (that is, at message 5 in Figure 48).

SSP-A receives an incoming call from a station belonging to another business group, or an outside line, or a station calling the attendant for service.

1 and 2   As no attendants are connected to SSP-A, the SSP requests the SCP to provide the address for the central attendant (or attendant group).

3   SSP-A forwards the attendant's network address to SSP-B with a request for connection.

4   SSP-B, to which the attendant is connected answers negative because the attendant is off-duty. The REL message contains an alternative network address of the night service station of that business group (Redirection Party Address).

5   SSP-A acknowledges the REL with RLC.

6   SSP-A makes another attempt to connect the call, this time using the night service station address at SSP-C.

Steps 4 and 5 may be repeated if the night service station (or MLHG) is busy. SSP-C would answer as SSP-C in step 4 and provide an alternative address. In this way, a MLHG or night service stations with series completion may be established across several exchanges.

SCP - SMS

No information is available about AWC specific messages sent between the SCP and the SMS. Transactions between these components are required for updating, reporting, statistics, and status information.

13.7 Traffic Measurement Requirements

Measurements must be provided, for example:
- numbers (of stations, attendants and MLHGs)
- group of numbers (for example to show traffic between two departments).

The items measured can include:
- peg counts
- response times
- congestion.

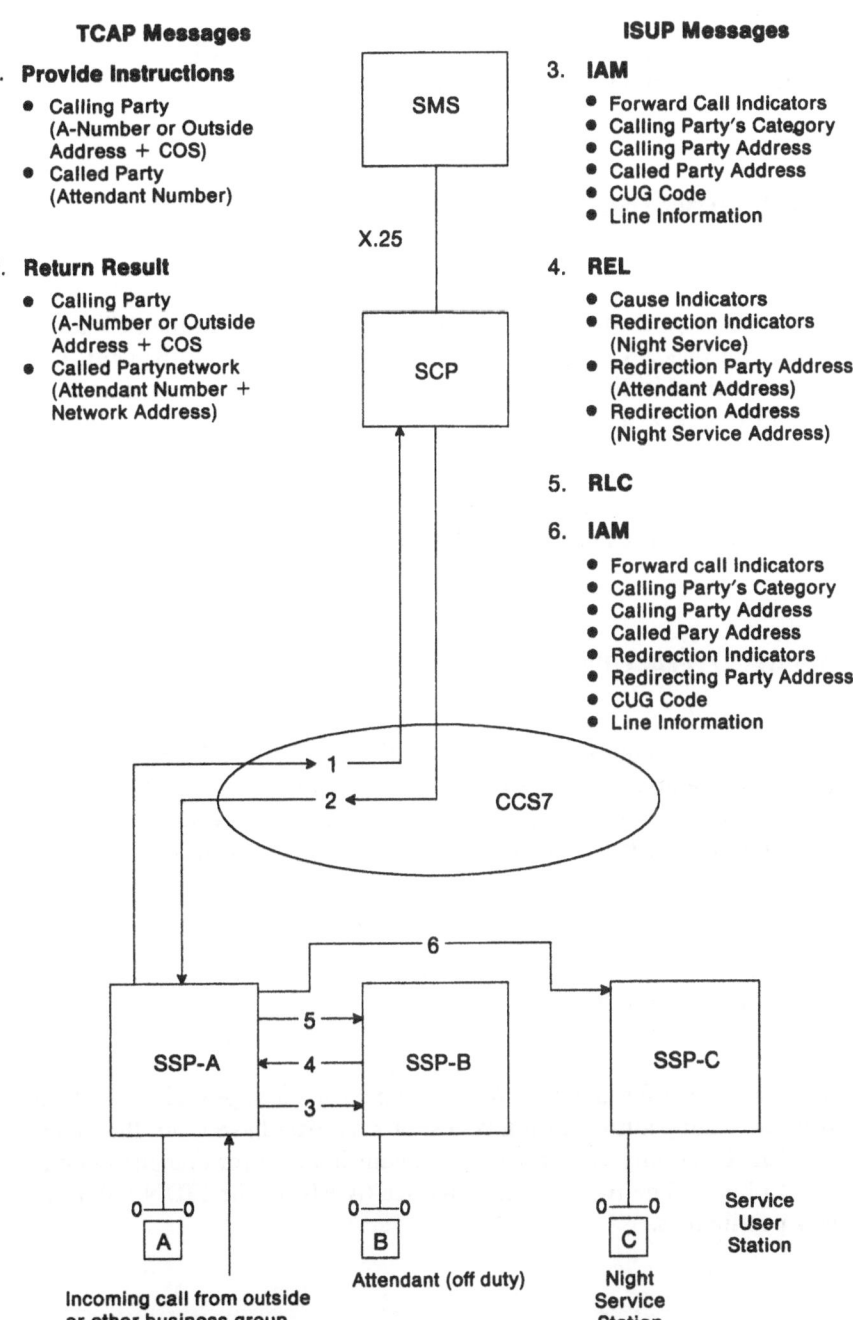

TCAP Messages

1. **Provide Instructions**
   - Calling Party
     (A-Number or Outside
     Address + COS)
   - Called Party
     (Attendant Number)

2. **Return Result**
   - Calling Party
     (A-Number or Outside
     Address + COS)
   - Called Partynetwork
     (Attendant Number +
     Network Address)

ISUP Messages

3. **IAM**
   - Forward Call Indicators
   - Calling Party's Category
   - Calling Party Address
   - Called Party Address
   - CUG Code
   - Line Information

4. **REL**
   - Cause Indicators
   - Redirection Indicators
     (Night Service)
   - Redirection Party Address
     (Attendant Address)
   - Redirection Address
     (Night Service Address)

5. **RLC**

6. **IAM**
   - Forward call Indicators
   - Calling Party's Category
   - Calling Party Address
   - Called Pary Address
   - Redirection Indicators
   - Redirecting Party Address
   - CUG Code
   - Line Information

SMS

X.25

SCP

CCS7

1
2

6
5
4
3

SSP-A   SSP-B   SSP-C

A   B   C   Service
                                  User
                                  Station

Incoming call from outside   Attendant (off duty)   Night
or other business group   Service
                                                      Station

Figure 48. Night Service Functional Flow

These measurements may vary greatly from country to country.

Generic traffic measurement requirements are described in Chapter 2, "Functional Characteristics Common to Selected IN Services".

13.8 Dynamic Requirements and Performance

SSP
The AWC user program should begin timing for a response from the SCP after sending a query message. A nominal value of three seconds should be used. If a response is not received in this period, a reorder tone or an announcement should be provided to the service user.

SCP
Upon receipt of a query from the SSP, the SCP should respond within 0.5 seconds on average and within 1.0 seconds 99% of the time.

13.9 National Dependencies

Like Centrex, AWC will compete with PBX systems. The introduction of AWC will depend on the country's PBX penetration and tariffs.

The degree of ISDN penetration is also important because ISDN contains most major functions of AWC. It will be up to the network operator to implement ISDN so that AWC can be offered as an alternative to PBX systems.

13.10 Future Considerations

New functions may be offered with AWC. Stations may change their assignment to a given group by dialing authorization codes, or may even have more than one assignment. This is not only valid for the attendant but also for company employees who work from home. The two channel interface (A + B) of the ISDN link also presents new possibilities.

Appendix A. Supplementary Services

1.   Call Waiting
2.   Call Transfer
3.   Called Line Identification
4.   Calling Line Identification Presentation
5.   Calling Line Identification Restriction
6.   Malicious Call Information
7.   Closed User Group
8.   Completion of Calls to Busy Subscribers
9.   Conference Calling
10.   Credit Card Calling
11.   Direct Dialing In
12.   DIVERSION SERVICES
12.1 Call Deflection
12.2 Call Forwarding Busy
12.3 Call Forwarding No Reply
12.4 Call Forwarding Unconditional
13.   Line Hunting
14.   Three Party Service
15.   User to User Signaling
16.   Multiple Subscriber Number
17.   Call Hold Service
18.   ISDN Networking Services
18.1 Private Numbering Plan
19.   Advice of Charge
20.   Reverse Charging
21.   Priority Service
22.   Terminal Portability
23.   Sub-Addressing
24.   Date and Time.

Appendix B. Bellcore Preliminary-Defined Functional Components and Requests[26]

A   Transfer of Control Functional Components
  1. Provide Instructions Request
  2. Release Request
B   Connection Functional Components
  1. Create Request
  2. Join Request
  3. Split Request
  4. Free Request
C   Network Participant Interaction Functional Components
  1. Send Request
  2. Receive Request
  3. Send & Receive Request
D   Network Information Transfer Functional Components
  1. Retrieve Information Request
E   Information Revision Functional Components
  1. Update Request
  2. Insert Request
  3. Delete Request
  4. Lock request
  5. Unlock Request
F   Processing Functional Components
  1. Monitor Request
  2. Cancel Request
G   Information Collection Functional Components
  1. Start-Collection Request

---

[26] Service Switching Point 2—SSP/2 Description, Bellcore SR-TSY-000782

Glossary

Note: Some of the terms are not directly related to IN, but have been included for completeness.

A-Link. Interface to the SS#7 Network

Appl. Application

AAR. Autonomous Analysis Reports

ABS. Alternate Billing Service

ACG. Automatic Call Gapping

ACH. Attempts per Circuit per Hour

AHP. Application host processor

AFR. Automatic Flexible Routing

ALI. Automatic line identification or automatic location identification. A database which provides street address information related to directory numbers.

AMA. Automatic Message Accounting

AMATPS. Automatic Message Accounting Tele Processing System

ANI. Automatic Number Identification. A local switching system capability to send to another switching system, the identity of a service user originating call.

ANSI. American National Standards Institute. National industry standards organization

API. Application program interface

APPC. Advanced program-to-program communications. An implementation of the SNA/SDLC LU6.2 protocol

APPC/PC. APPC program for the IBM Personal Computer

APS.
1. EWSD Application Program System
2. Application Service Part

ASD. Autonomous Surveillance Data

ASN. Abstract Syntax Notation

ASP.
1. Adjunct Service Point, a functional module of IN/2.
2. Application Service Part

AWC. Area Wide Centrex

BA. Basic access

Basic Services. A FCC-defined regulatory term. Basic services are limited to those providing a transparent communications path or pipeline. The provider can not control or manipulate the contents of the transmissions. Basic services include WATS, private line, video transmission services and services closely related to basic services such as abbreviated dialing, directory assistance, billing, and voice encryption services. The basic services are regulated by FCC[27].

Bellcore. Bell Communications Research. Research and development company jointly owned by the seven regional Bell operating companies

B:IOC. Bus Interface

B:CMY. Access Bus to Common Memory

BSDB. Business Service Database

BS2000. Siemens virtual memory operating system for 7.5XX series computers

BX.25. Bell communications protocol standard (used by the Operations System Network)

CASM. Communication And System Management (PS/2 SW)

CB. Channel Bank

---

[27] From IEEE Communications Magazine August 1987, p36.

CCG. Central Clock Generator
CCIS. Common channel interoffice signaling
CCITT. International Telegraph and Telephone Consultative Committee. International standards organization
CCNC. Common Channel Signaling Network Control of EWSD
CCNP. Common Channel Signaling Network Processor of EWSD
CCS. Common Channel Signaling
CCS7. Common channel signaling system No. 7, defined by CCITT
CCSN. Common Channel Signaling Network. Uses the CCS7 communications protocol
CCV. Calling card validation
CEO. Chief Executive Officer
CHILL. Programming language standardized by CCITT
CI. Common Interface
CICS. Customer Information Control System. IBM's most widely used Teleprocessing Monitor providing application programs with terminal and database support.
CLASS. Customer local area signaling services. A group of services such as:
- Automatic Callback
- Automatic Recall
- Calling Number Delivery
- Calling Number Delivery Blocking
- Customer Originated Trace
- Distinctive Ringing/Call Waiting Screening List
- Selective Call Forwarding
- Selective Call Forwarding Screening List
- Selective Call Rejection
- Selective Call Rejection Screening List
- Who Called Me.

CLLI. Common Language Location Identifier
CMSDB. Call-Management-Services Database
CMY. Common Memory
CO. Central Office. Switching node nearest to the subscriber
CP. Coordination Processor

CPA. Called or Calling Party Address field of the CCS7 protocol
CPE. Customer Premises Equipment
CPR. Call Processing Record
CSS. Customer service system. A Telco system applied in administration activities to store and process telephone subscriber and service order related data.
CST. Call setup time
DAS. Directory assistance service. A service providing a telephone number to a person who knows the name and address of the party to be called.
DASD. Direct Access Storage Device. Secondary (disk) storage
Database machine. A specialized processing system for database management. Often called a back-end processor. It consists of processor and storage hardware, and data base and operating system software. It is designed to handle the data storage and retrieval functions for a host CPU. Software relational database systems occupy a lot of host resources, resulting in a rather slow response time. Database machines allow greater processing speed, especially on large, complex databases.

Most database machines support the relational database model. Some vendors:
Britton Lee - IDM
Teradata - DBC/1012.
DB. Database
DB2 - Database 2. IBM's newest relational database management system. Runs on IBM/370 mainframes.
DBP. Deutsche Bundespost
DBS. Database Service
DSAC. Dial Service Administration Center
DCE. Data circuit-terminating equipment of the X.25 protocol
DCN. Data Communication Network
DFD. Data Flow Diagram
DLU. Digital Line Unit
DLUC. Digital Line Unit Card
DMS. Data management system. The DMS is a system of manual procedures and computer programs used to create, store, and update data required for selective routing.

DOC. Dynamic Overload Control
DOD. Direct Outward Dialing
DPC. Destination Point Code (of the CCS7 protocol)
DPN. Dedicated Private Network
DSAC. Dial Service Administration Center
DSNX. Distributed Systems Node Executive (SW)
DSX. Distributed Systems Executive (SW)
DTE. Data Terminal Equipment (of the X.25 protocol)
DTMF. Dual Tone Multifrequency
EADAS. Engineering and Administration Data Acquisition System
EMML. Extended Man Machine Language
Enhanced Services. An FCC-defined regulatory term. Those services where the provider employs a computer to act on the "format, content, code, protocol, or similar aspects of the subscriber's transmitted information; provide the subscriber additional, different, or restructured information; or involves subscriber interaction with stored information[28].

An enhanced service usually involves an underlying basic service offering (such as using some carrier's terrestrial or satellite switched or private circuits) on which a computer is used to manipulate the code or content of the transmissions.

Examples of enhanced services include voice store-and-forward services (and other services which emulate telephone answering machines), linking services which permit incompatible computers to communicate with each other, and mass calling services, such as dial-up stock quotation or sports services. The enhanced services are not regulated by FCC.
ERS. Emergency Response Service
ESDS. Entry Sequence Data Storage method of VSAM
ESP. Enhanced service provider. An entity that provides enhanced services to the network.

EWSD. Elektronisches Wähl System Digital, a Siemens Digital Switch
FAR. Facility Request Message
FC. Functional component. Basic service building block such as JOIN which can connect a line of to an existing call, or CREATE which creates a connection from one responding station to another.
FCC. Federal Communication Commission
FISU. Fill-in signal unit of the CCS7 protocol
FRJ. Facility Rejected Message
FSD. Feature Specific Document
FT. File Transfer
FTAM. File Transfer, Access and Management (an OSI application Layer service)
GNS. Green Number Service
GP. Group Processor
GTT. Global Title Translation
HMI. Human Machine Interface
Hook Flash. The brief signal received when a user manually presses the hook switch or hand set switch, as if returning the receiver to an on-hook state.
HT. Holding Time
HTR. Hard-To-Reach
HU. High Usage
IC. Interexchange Carrier (or IEC)
ID. Identification
IDC. Information Distribution Company
IMA. Ineffective Machine Attempts
IMS. Information Management System. IBM's most widely used database management system.
IN. Intelligent Network
INC. International Exchange
IN/1. Intelligent Network Phase 1. Supports simpler services such as the 800 service and calling card verification.
IN/2. Intelligent Network Phase 2. Planned to support FCs operating in SCPs, SSPs, or IPs.
IOC. Input/Output Controller
IOP. Input/Output Processor
IP. Intelligent Peripheral. Provides enhanced services under the control of an

---

[28] From IEEE Communications Magazine August 1987, p36.

SCP or SSP. The services provided are not normally in the SSP, or it is too expensive to put in all SSPs. Thus, it is more economical to share an IP between users.

IPL. Initial program load

ISDN. Integrated Services Digital Network

ISDN Island. An introduction strategy for ISDN. ISDN service users are connected to the same exchange, and cannot communicate with ISDN users on other ISDN exchanges due to lack of CCS7 connections between the exchanges. The linking of the ISDN exchanges via CCS7 eventually enables all ISDN subscribers to communicate with each other.

ISO. International Standards Organization

JES. Job entry subsystem

KDCS. Kompatible Daten Kommunikations Schnittstelle (Compatible Data Communication Interface). A standard program interface to TP monitors.

KSDS. Keyed Sequential Data Storage method of VSAM.

LAN. Local area network. In the SCP it refers to the token ring and attachments to it in the SCP.

LAPB. Link Access Procedure Balanced

LATA. Local Access and Transport Area

LCD. Liquid-Crystal Displays

LEC. Local Exchange Carrier

LIDB. Line Information Database. Makes the subscriber data (profiles) widely available to the switching systems and thus allows increased portability of customer services. The centralized customer data also facilitates automation of services offered by Telcos on a manual basis.

The LIDB will initially validate calling cards validation, screen originating line numbers, select carriers and control fraud. The LIDB does not apply in every country, due to different national and legal situations.

LMY. Local Memory

LSSGR. LATA Switching Systems Generic Requirements

LSSU. Link Status Signal Units

LSTP. Local Signaling Transfer Point

LTG. Line/Trunk Group

LU. Logical unit (of the SNA protocol)

LU6.2. Logical Unit Type 6.2 (of the SNA protocol) Supports distributed transaction processing at SNA network nodes

MB. Message Buffer

MC1. Machine Congestion, Level 1

MC2. Machine Congestion, Level 2

MF. Multi Frequency Signaling. MF signaling arrangements use pairs of frequencies from the six frequencies: 700, 900, 1100, 1300, 1500, and 1700 Hz. MF signals are used for called number address signaling, calling number identification, ringback, and coin control. The possible frequency combinations represent the digits 0 through 9 and many special control or information signals.

MML. Man-machine language. Used to define human interfaces

MSU. Message signal unit of the CCS7 protocol

MTP. Message Transfer Part. Levels 1-3 of the CCS7 protocol

MVS. Multiple virtual storage. IBM's operating system for the /370 series computers.

MVS/XA. Multiple Virtual Storage/Extended Architecture

MYB. Memory Bank

MYC. Memory Control

NAND. Not AND (Logical Operator)

NBS. National Bureau of Standards

NCCS. Network Control Center System

NCSC. National Computer Security Center

NDS. Network Data System

NE. Network Element

NID. Network Information Database, IN/2

NMA. Network Monitoring and Analysis

NOR. Not OR (Logical Operator)

NPA. Numbering Plan Area (Area Code)

NPMS. Network Performance Management System

NRM. Network Resource Manager, IN/2

NSEP. National Security Emergency Preparedness

NSMA. Network and Service Management Architecture

NSMS. Network and Service Management System

NTE. Network Terminating Equipment

NTM. Network Traffic Management

NXX. End-Office Code

OA&M. Operation, Administration and Maintenance

OCCF. VSE/Operator Communication Control Facility IBM licensed program to centralize operator interaction

OLNS. Originating Line Number Screening

OLTP. Online transaction processing. Supports high reliability (fault tolerant) online database transactions. Typical examples are Automatic Teller Machines and airline reservation systems.

OMAP. Operations, Administration and Maintenance Application Part

OMDS. Siemens Operation and Maintenance Data Communications System

ON. Open Network. A technological concept to describe a network which can interconnect system elements provided by vendors whose designs meet interface standards.

ONA. Open Network Architecture. A regulatory concept promulgated by the FCC which requires the Bell Operating Companies to offer access to basic service elements (BSEs) for the provision of enhanced services if they want to offer enhanced services without structural separation.

OPC. Originating Point Code (of the CCS7 protocol)

OPDU. Operation Protocol Data Unit

OS/2. Operating System/2 for the IBM PS/2

OSI. Open System Interconnection

OSN. Operations System Network (uses the BX.25 communications protocol)

OSNI. Operations System Network interface, front end processor which provides the SCP interface to the BX.25 operations support network

OSS.
1. Operator Services System
2. Operations Support System

OSSGR. Operator Support System Generic Requirements

OT. Operations Terminal

OTGR. Operations Technology Generic Requirements

PA. Primary access

PC. Personal computer. This document usually refers to the PS/2 model(s)

PCM. Pulse Code Modulation

PDC. Primary Digital Carrier

PDM. Platform Data Manager of an application host

PIN. Personal Identification Number

PPS. Public Phone Service

PSPDN. Packet Switched Public Data Network

PS/2. Personal System/2 IBM product

PSAP. Public Safety Answering Point. An agency or facility designated by a municipality to receive and handle emergency calls. A PSAP may be designated as primary or secondary, which refers to the order in which calls are directed. Primary PSAPs receive calls on a transfer basis only. PSAPs have also been generally referred to as emergency service bureaus (ESBs).

PSPDN. Packet Switched Public Data Network

PSTN. Public switched telephone network

PTT. Post, Telephone, and Telegraph Company

PU.
1. Physical unit of the SNA protocol
2. Processing Unit

PU T2.1. Physical Unit Type 2.1 used by PCs to support distributed transaction processing in an SNA network

PU T5. Physical Unit Type 5 used by IBM System/370 hosts in an SNA network

PVN. Private Virtual Network; A service with which a business can tailor its own network based on the resources of the RBOCs and ICs. The user pays for the resources, and Telcos guarantee that the resources will be there when needed, without actually allocating a physical resource permanently to a specific user. Often the PVN is offered with a network management system that allows the user to manage his own PVN.

RAO. Revenue accounting office

RBOC. Regional Bell Operating Company.

RDBMS. Relational Database Management System; A database management system built on the relational model. All data resides in tables, and the user can use a standardized database language (SQL) to manipulate the data.

RDBMS hides the data structures and the connections between data for the user, allowing him to formulate queries strictly with logical expressions (non-navigational), for example:

Car Color = RED   and
Car Type = MERCEDES and
Location = MUNICH

gives a result of all red Mercedes cars in Munich.

RIPL. Remote Initial Program Load

RMAS. Remote Memory Administration System

RNR. Receiver-Not-Ready

RR. Reroute control

RSTR. Regional Signaling Transfer Point

RS-232-C. An asynchronous protocol for serial interfaces used to transmit digital information over short distances

SAA. IBM Systems Application Architecture

SAC.
1. Service administration center
2. Service access code

SCCP. Signal Connection Control Part, level 4 of the CCS7 protocol

SCCS.
1. Bellcore's Service Control Center System
2. Switching Control Center System

SCP. Service Control Point at which a new service is normally introduced. If a service is executed based on functional components, (IN/2), the FCs required are executed using a service logic interpreter using a script that explains how the FCs should be executed.

Some services in the SCP may require a large amount of data which must then reside on disks.

The service programs and the data are updated from the SMS. The SCP can be either a modified switch or a commercial mainframe.

SDC. Secondary Digital Carrier (8M bps)

SDLC. Synchronous data link control communications protocol

SEAC. Signaling Engineering Administration Center (for CCS7)

SEAS. Signaling, Engineering, and Administration System

SGC. Switch Group Control

SHIP. SPI Host Interface Programs

SI. Service Indicator field of the CCS7 protocol

SILC. Selective Incoming Load Control

SIO. Service Information Octet field of the CCS7 protocol

SLC. Signaling Link Code, identification of a signaling link within a Signaling Link Set

SLI. Service Logic Interpreter; a functional module of IN/2. The SLI interprets service-logic programs that have been defined by a Telco or a customer.

SLP. Service Logic Program, a functional module of IN/2

SLS. Signaling Link Selection field of the CCS7 protocol

SMS. Service Management System. The SMS provides the service subscriber with the ability to control his own service via a terminal linked to the SMS. The SMS is responsible for updating all SCPs with new data or programs, and collecting statistics from the SCPs. The SMS is usually a commercial mainframe, such as Siemens 7.5xx or IBM/370.

SN. Switching Network

SNA. Systems Network Architecture, IBM communications protocol

SNM. Signaling Network Management (function of the CCS7 protocol)

SOCC. Selective Originating Call Control

SP. Service Provider

SPC. Signaling Point Code. Address of a signaling point in a CCSN

SPCS. Stored Program Control System

SPI. Signal Point Interface. Front end processor that provides the SCP interface to the CCS7 signaling network

SQL. Structured Query Language. Database language based upon the

SEQUEL language. The user can manipulate a relational database with no knowledge of the physical implementation of the database (non-navigational).

SR. Special Report

SSA. Structured System Analysis

SSCP. System Services Control Point of the SNA protocol which manages the resources of a Type 5 SNA node

SSN. Subsystem Number of the CCS7 protocol used to identify an application

SSP. Service Switching Point. Access point for service users and executes heavily used services.

The SSP is produced by traditional switch manufacturers.

SSP/2. SSP for IN/2

CCS7. Signaling System Number 7 communications protocol standard defined by the CCITT and ANSI standards organization (used by the Common Channel Signaling Network)

STP. Signaling Transfer Point. Part of the Common Channel Signaling No. 7 (CCS7) network, it is responsible for switching CCS7 messages between different CCS7 nodes.

The STP is normally produced by traditional switch manufactures.

SU. Signal unit, a unit of information defined by the CCS7 protocol

SV. Service Vendor

SYP. System Panel

TA.
1. Technical Advisory
2. Transaction Age

TC. Transaction Component

TCAP. Transaction Capabilities Application Part (level of the CCS7 protocol)

Telco. Telecommunication Company

TL1. Transaction language #1 (based on CCITT MML standard)

TMN. Telecommunication Management Network

TP. Technical Paper

TPS. Transactions per second

TR. Technical Requirements

Traveling Classmark. The final one or two digits of an incoming tie-trunk call determining the user's calling characteristics and limitations.

UA. User Access

UAL. User Application Layer (of the BX.25 protocol)

User Programmability. The ability of Telcos to program a service themselves, or via a software house, and introduce this service into the network.

UTM. Universal transaction monitor; A BS2000 TP-Monitor.

V.35. CCITT protocol used to transmit data across CCS7 signaling links

VC. Virtual Circuit

VCA. Vacant Code Announcement

VFN. Vendor Feature Node outside the network. Owned and administered by a service supplier. It provides many services.

VMS. Virtual Memory System

VSAM. Virtual Storage Access Method for disk data access

VSE. Disk Operating System/Virtual Storage Extended

VTAM. Virtual Telecommunications Access Method

X.25. CCITT standard protocol for packet information transfer (used by the Public Packet-Switched Network)

1TR7. Application Specification for CCITT Signaling System No. 7 in the national network of the German Bundespost

1/1AESS. No. 1/1A Electronic Switching System

4ESS. No. 4 Electronic Switching System

5ESS. No. 5 Electronic Switching System

Bibliography

For information regarding the purchase of Bellcore copyrighted documents contact:

Bellcore Customer Service
60 New England Avenue
Piscataway
New Jersey 08854-4196
USA

Additional Service Switching Point Capabilities (Including Private Virtual Network Services), Bellcore TA-TSY-000402
Automatic Callback, Bellcore TA-TSY-000215
Automatic Recall, Bellcore TA-TSY-000227
Basic Access Switching and Access Requirements, Bellcore TA-TSY-000268.
Bulk Calling Line Identification, Bellcore TA-TSY-000032
Business Services Database (BSDB): SCP Applications Designed to Support Private Virtual Network (PVN) Services, Bellcore TA-TSY-000460
Calling Number Delivery, Bellcore TA-TSY-000031
Calling Number Delivery, Bellcore TR-TSY-000319
Common Channel Switching System Maintenance Requirements, Bellcore TA-TSY-000372.
Customer Originated Trace, Bellcore TR-TSY-000216
Distinctive Ringing/Call Waiting, Bellcore TR-TSY-000219
Exchange Access Alternate Billing Services, Bellcore TA-TSY-000400

Exchange Alternate Billing Services, Bellcore TA-TSY-000399
E911 Public Safety Answering Point: Interface between a 1/1AESS Centrex Office and Customer Premises Equipment, Bellcore TA-TSY-000350
Feature Transparency for Multi-Location Business Users, Bellcore TA-TSY-000404
FSD 31-01-0000 TCAP Message Format for 800 Service, Bellcore TR-TSY-000064
LSSGR LATA Switching Systems Generic Requirements, Bellcore TR-TSY-000064
Network Traffic Management (NTM) Operation System (OSS) Requirements, Bellcore TA-TSY-000753
Plan for the Second Generation of the Intelligent Network, Bellcore SR-NPL-000444.
Reliability and Quality Switching Systems Generic Requirement (RQSSGR), TA-TSY-000284.
RFI - Feature Node/Service Interface Concept, Ameritech, Issue1, February 1985, Bellcore SR-NPL-000108,
Screening List, Bellcore TA-TSY-000220
Selective Call Forwarding, Bellcore TR-TSY-000217
Selective Call Rejection, Bellcore TR-TSY-000218
Service Control Point Node - Generic Requirements, Bellcore TA-TSY-000029
Service Logic Interpreter Preliminary Description, Bellcore SR-TSY-000778
Service Switching Point 2 (SSP/2) Description, Bellcore SR-TSY-000782
Service Switching Points, Bellcore TR-TSY-000064, FSD 31-01-0000.

Service Switching Point 2 - SSP/2 Description, Bellcore SR-TSY-000782.
Signaling Transfer Points Generic Requirements, Bellcore TR-TSY-000082.
Support System Interface Protocol, Bellcore TR-TSY-000510 (part of LSSGR TR-TSY-000064)
SCP-SMS Generic Interface Specification, Bellcore TA-TSY-000365.
SMS/LIDB-LIDB Interface Specification, Bellcore TA-TSY-000446
SMS/800 - Terminal Data Communications Planning Information, Bellcore SR-STS-000742.
800 Service Management System User Guide: General Procedures, Bellcore SR-STS-000741
800 Service Management System User Guide: 800 Number Administration, Bellcore SR-STS-000740
The Geodesic Network - 1987 Report on Competition in the Telephone Industry, Peter W. Huber
CCITT TCAP preliminary Blue Book
AT&T's Pay-per-View Television Trial, AT&T Technical Journal, Volume 66, Issue 3
ISPF - Dialog Management Services, SC34-2088-1 (IBM Publication)
Freephone Supplementary Service procedures
CEPT, Sub-working group SPS/PAR, October 1987
Changes in Observable Call Setup Time with Network Evolution Gale McNamara, Bell Communications Research, from National Communications Forum 1987
Various articles from:
1. IEEE Communications Magazine
2. Telecommunications
3. Bellcore IN workshop material
4. IBM Systems Journal
5. Data Communications
6. Telephony
O&M-Concepts for Public Networks using ISDN, CCS7 and PCs, Siemens, 1987
IEEE 1988 Network Operations and Management Symposium (NOMS) Proceedings February 29, 1988
K. C. O'Brien, *Evolving Operations Architectures*,
A look to the future, Bell Communications Research, Inc.
H. P. Brandt, E. Eastland, *Network Operations Architecture for ONA*, Northern Telecom, Inc., Bell Northern Research
H. B. Gangele, Datenkommunikationssystem OMDS unterstützt "maßgeschneiderte" Vermittlung, Siemens telcom report 10 (1987) Nr. 3
R. Fröschle, M. Kay, *Alarm Signaling in the EWSD Digital Switching* Siemens telcom report 10 (1987) Nr. 4
L.R. Bowyer, M.L. Almquist, GLOBECOM '86: IEEE Global Telecommunications Conference, Communications Broadening Technology Horizons.
Conference Record (Cat. No.86CH2298-9) Houston Texas, USA 1-4 Dec. 1986 p.1325-9 vol.3 1986
L.R. Bowyer, M.L. Almquist, The Signaling Engineering and Administration System (Article) Bell Communication Research. 444 Hoes Lane, Piscataway, New Jersey 08854